How do you get the best from your telescope? For those interested in photographing the Sun, Moon and planets, this volume provides the complete reference. This guide is packed with practical tips on how to obtain the highest resolution in your astrophotography, and provides a wealth of stunning images by the world's best amateurs showing just what can be achieved.

Individual chapters describe the various types of telescopes and the most suitable equipment to photograph a given subject, and recommend films and techniques in developing and printing. Also given are short biographies of key high resolution astro-photographers, both past and present, and an extensive bibliography of further reading.

This guide provides both a wealth of sound, practical techniques and a unique portfolio of solar system images – an inspirational handbook for any amateur astronomer.

High Resolution Astrophotography

Practical astronomy handbook series

The Practical Astronomy Handbooks are a new concept in publishing for amateur and leisure astronomy. These books are for active amateurs who want to get the very best out of their telescopes and who want to make productive observations and new discoveries. The emphasis is strongly practical: what equipment is needed, how to use it, what to observe, and how to record observations in a way that will be useful to others. Each title in the series will be devoted either to the techniques used for a particular class of object, for example observing the Moon or variable stars, or to the application of a technique, for example the use of a new detector, to amateur astronomy in general. The series will build into an indispensable library of practical information for all active observers.

Titles available in this series

1. A Portfolio of Lunar Drawings
   by Harold Hill
2. Messier's Nebulae and Star Clusters
   by Kenneth Glyn Jones
3. Observing the Sun
   by Peter O. Taylor
4. The Observer's Guide to Astronomy (Volumes 1 & 2)
   edited by Patrick Martinez
5. Observing Comets, Meteors, and the Zodiacal Light
   by Stephen J. Edberg and David H. Levy
6. The Giant Planet Jupiter
   by John H. Rogers
7. High Resolution Astrophotography
   by Jean Dragesco

High Resolution Astrophotography

Jean Dragesco

translated by and with the collaboration of
Richard McKim

CAMBRIDGE
UNIVERSITY PRESS

Published by the Press Syndicate of the University of Cambridge
The Pitt Building, Trumpington Street, Cambridge CB2 1RP
40 West 20th Street, New York, NY 10011-4211, USA
10 Stamford Road, Oakleigh, Melbourne 3166, Australia

Printed in Great Britain at the University Press, Cambridge

A catalogue record for this book is available from the British Library

Library of Congress cataloguing in publication data available

ISBN 0 521 41588 8 hardback

Contents

Preface

Telescopic planetary photography is an art. You must have talent. For Jean Dragesco it has also been a calling.

Formerly professor at the Federal University of Cameroun, researcher of the French scientific organisation CNRS,[†] zoologist and film-maker, Jean Dragesco personifies protistology. For him, the little creatures which are called Ciliates are the subject of his research, the passion of his work.

Jean Dragesco also incarnates Africa. Mauritania, Chad, Gabon, Kenya, Tanzania, Benin and Rwanda have been his places of study: a life of field and laboratory work which earned him fine scientific achievements as well as his celebrated films on the birds and mammals of the great African continent.

But while he headed the general Biology Laboratory at Yaoundé and was teaching in the City University, Jean Dragesco, as everywhere else, had the habit of mysteriously disappearing during the night. In his garden or on his terrace, weird lights could then be perceived, under the wonderfully starry celestial vault. The outlines of a strange cylinder seemed to be pointing to a region of the sky, while a ghost moved slowly about behind it. Some noises, too: a constant humming and strange clicking sounds. In the direction of the cylinder, amidst the celestial vault, a star shining more brightly than the others. The strange stillness and mystery of the African night. The impression of a serious and important operation.

Jean Dragesco has owned 32 telescopes. At each of his residences, which have been numerous, he has installed an observatory; planetary observing was obligatory. One can hardly imagine Jean Dragesco without a telescope. Chopin without a piano? . . . a question of temperament.

Jean Dragesco is also an apostle: he converts others to his cause. He excels in bringing out the first fruits of talent, in the youngest people, in astronomical observation. He supervises them,

† Centre National de la Recherche Scientifique.

guides them, values them. I had the proof when I made his acquaintance in the SAF (Société Astronomique de France), in 1941. He was my senior. Recognising what was evidently a promising subject for his very young friend (did he have a flair for it?), he immediately took me under his wing.

It was he who instructed me in the polishing of the mirror, the principal part of the telescope I wished to build. It was he who lent me a small 75 mm refractor, with which I observed Saturn and the Moon. Then he lent me a 135 mm objective which I mounted in a tube that I had for my first refractor. It was with him again that I observed the planet Mars for the first time with a real equatorial telescope, on 1941 July 14 at 18h 15m at the SAF observatory. Since then I have often returned to view Mars with a refractor.

The SAF has always been a nursery for amateur astronomers, volunteers and patrons. There, one can find potential observers and astrophotographers. In 1956, at the time of the close approach of the planet Mars to the Earth, a group was constituted within the SAF called the 'Commission des Surfaces Planétaires'. If I had to coordinate it myself in the first place, it was Jean Dragesco who was really the moving force. Afterwards, he succeeded me in this role, taking everything in hand, and improving things which, owing to circumstances, I had to skate over.

For us all, in our group, it was a treat to see Jean Dragesco working alongside and for the younger generations of devoted astronomical observers. He doesn't surprise us, in the evening of his life, by giving us once again the benefit of his knowledge in 'High Resolution Astrophotography'.

The work is a rich one. It would have been easy for Jean Dragesco to have merely drawn upon his unique personal documentation, to have given just his own experiences. The illustrations could have been drawn entirely from the remarkable photographs that he himself had obtained. But his sense of fairness came first, in a way which offers the reader a much more elaborate analysis, being based upon the best expertise and the most superb results available. He searched everywhere for the best talents, identified where they could be found, set down their experiences and presented their results in his book in just recognition of them and to the benefit of the reader.

The result is astonishing, since the quality of the planetary images presented is unique. There, one can appreciate the high level that amateur astronomers can reach through their passion, artistry, talent, perseverance. Frequently in this field, their productivity largely exceeds that of the professional observatories, geared towards other fields of study.

This book has attained its goal. It reveals a desire for telescopic observation, to practise planetary photography, to perfect the art, to begin with interesting photographs, and then to systematically produce remarkable images. It thus invites us to contribute to the surveillance of planetary surfaces and atmospheres, a necessity for Science, in which amateur astronomers have imposed themselves as masters of the art.

Audouin Dollfus
Astronome titulaire de l'Observatoire de Paris

Acknowledgements

I owe my early interest in Astronomy to my uncle Dr Radu Drăgescu, who introduced me to the night sky and guided my reading towards the works of Camille Flammarion, and later to those of the Abbé Th. Moreux and Lucien Rudaux. It was with my colleague, C. Dobrovici, that I was able to found in Bucharest, in 1938, an association of young amateur astronomers, and to edit a monthly review, *Urania*.

From the beginning of my time in France, from 1940 onwards, the Société Astronomique de France helped me a good deal, and it was thanks to Madame Gabrielle Flammarion that I had access to the two domes of the SAF observatory. I was thus able to begin serious planetary observing under the guidance of G. Fournier and G. de Vaucouleurs. A little later, I was able to collaborate with G. de Vaucouleurs in the field of scientific photography. Finally, it was thanks to his influence that I had the opportunity to work at the Lowell Observatory at Flagstaff, where I was very well received.

It was at the observatory of the SAF that I got to know Audouin Dollfus, when he was but a young man. Later, Professor Dollfus was to play a large part in my life: he gave me the chance to direct the *Commission des Surfaces Planétaires* of the SAF, and later introduced me to the Centre for Planetary Documentation at Meudon, where I had access to the Grande Lunette of 83 cm aperture. It was, again, Audouin who recommended me to the Directors of the Pic du Midi Observatory.

Like all active French amateurs, I was to learn a great deal from Jean Texereau, at the Laboratoire d'Optique of the Paris Observatory, at the SAF headquarters, or through reading his monthly contributions to the SAF bulletin, *l'Astronomie*. Texereau taught us, above all, both critical spirit and humility.

Charles Boyer helped me a good deal too. I met him at Meudon, as well as at the Pic du Midi Observatory, where I had a friendly reception from Professor Jean Rösch.

The Société Astronomique de France, like the British Astronomical Association and the Association of Lunar and Planetary Observers (USA), has assisted me in the fields of planetary and solar observations, thanks to the following friends and colleagues: Régis Néel, Jean Dijon, R. Gili, René Verseau, Dr Christian Botton, J. Lecacheux in France; E. H. Collinson, W. E. Fox, Dr J. H. Rogers and Dr R. J. McKim in England; M. Gomez, L. Tomàs, A. Sanchez-Lavega in Spain; Dr Donald C. Parker, J. D. Beish, P. W. Budine, P. K. Mackal, Dr J. L. Benton, Dr J. E. Westfall, R. Hill, P. O. Taylor in the USA; Prof. S. Miyamoto, T. Osawa, T. Saheki, T. Akutsu in Japan; and Matei Alexescu and D. Vidican in Romania.

Georges Viscardy was a great friend to me, both as an example and as a guide. I must not forget, either, the wonderful moments spent, with the 106 cm telescope of Pic du Midi, with Dr Richard McKim, who has helped me so much with his advice, and who translated the present work. I thank Dr Simon Mitton, who gave me the chance to publish this book, and also all those who sent me their superb photographs and precious information, notably: C. Arsidi, A. Behrend, J. Deconihout, B. Flach-Wilken, Cord-Hinrich Jahn, F. Küffer, D. Lachaud, W. Lille, I. Miyazaki, G. Nemec, Dr D. C. Parker, T. Platt, Prof F. Rouvière, G. Thérin, L. Tomàs and G. Viscardy.

To conclude, I pay homage here to those intertropical African countries, where I spent 20 years of my life and which allowed me to experience such magnificent skies. Above all, I thank my wife Armelle, who reread some hundreds of pages of manuscript, greatly improving the text.

Jean Dragesco
St. Clément-de-Rivière

Introduction

Amateur astrophotography has several facets. The field of high resolution photography seems, to the majority of astrophotographers, to be the most difficult, best left to a small number of experts. The term 'high resolution' is ambiguous, for it is open to two very different interpretations of the problem that interests us: on the one hand, absolute high resolution, and, on the other, relative high resolution. In principle, a high resolution photograph is one which shows us the finest details of a given celestial body, independent of the instrument used. One such concept is much too large, as it is evident that there are spatial factors which prevail; upon the Moon, as upon Mars, the best resolution is of the order of a few millimetres (photographed by the Apollos and the Vikings). We admit, therefore, that we are only talking about photographs taken with Earth-based telescopes when we speak of high resolution. Thus, in high resolution work we seek to record by photography the finest details of the celestial bodies we can see with our telescopes. The resolution is defined in an angular manner, in seconds of a degree of arc ($''$) (rather than in kilometres or centimetres on the surface of the body in question).

Theoretically, the largest telescopes should allow us to reach the highest resolution (since we know that the resolving power is directly proportional to the diameter of the objective used). In practice, however, the Earth's atmosphere limits the effective resolving power; this limitation is very severe and offsets the advantage of the largest instruments (with the exception of the Hubble Space Telescope, which, being situated outside the atmosphere, is able to utilise fully the theoretical resolving power of its 2.4 m mirror, at least now that its spherical aberration has been completely corrected).

For this reason amateurs find themselves in a privileged position: their instruments, of much smaller diameter, are less severely affected by atmospheric turbulence (seeing). With skilful application, being both ingenious and patient, amateurs are successfully photographing details on the bodies of the Solar System as fine as those recorded by some of the most powerful telescopes (handicapped by their over-large apertures). However, this situation is temporary, because the large telescopes will soon be able to correct for the effects of turbulence, thanks to their adaptive optics, and amateurs will be rapidly left behind.

We can also consider high resolution astrophotography in another manner (Dragesco, 1978), not in absolute terms but as a function of the theoretical limits of the instrument used. Thus, a photograph with resolution reaching $1''$ obtained with a 15 cm telescope will be considered as a high resolution one, whereas another, which reaches $0''.5$ but which is obtained with a 100 cm telescope, will represent a much less satisfactory performance. It follows that high resolution is for the taking by everyone: it is, very simply, a question of attempting to reach the limits of a given instrument, by skilful application, experience and perseverance. A small-diameter telescope will therefore be favoured, since it will be less susceptible to atmospheric turbulence. The owner of a small 60 mm refractor will very easily be able to obtain photographs reaching maximum resolution, which the instrument warrants, whereas an observer using a 600 mm telescope will probably never be able to reach the resolution limit.

Thus, it is less frustrating to be content with an excellent telescope, of average size (200–250 mm), whose theoretical resolving power is good enough to compete with some of the largest telescopes. However, that requires a rare virtue, the mastering of difficult techniques and extreme persistence. As we shall see, some amateurs have succeeded, with considerable difficulty, in obtaining high resolution photographs.

High resolution photography for amateurs has made tremendous progress in the last few years, and it will suffice to compare the figures published by Saget (1952), and by Texereau and de Vaucouleurs (1954), with those which illustrate this book.

Initially progress was very slow, even for professional astronomers. According to de Vaucouleurs (1958), the adventure began between 1890 and 1900, with the first good photographs of the Moon, obtained at the observatories of Lick (91 cm refractor) and Paris (60 cm refractor). However, the resolution remained very low: the finest details recorded were easily visible in the eyepiece of a small 6 cm or 9 cm refractor (the 'efficiency' was therefore not more than a mere 1/10). The details on the lunar photographs obtained on Mount Wilson with the 2.5 m telescope, in 1920, were easily visible in an amateur-sized telescope of 20 cm aperture (efficiency only 1/12: the telescope was decidedly too big!). From 1905 to 1930 at the Flagstaff Observatory, Lowell, Lampland and Slipher, with the help of the 60 cm Clark refractor (diaphragmed, most usually to 45 cm), succeeded (rarely) in photographing planetary details as small as 0".6, corresponding to an efficiency reaching 1/4.

The modern period of high resolution photography began about 1941 with Lyot, Camichel and Gentili who, with the 30 cm and 60 cm refractors at Pic du Midi, sometimes recorded planetary details as small as 0".6 or 0".4 (resulting in an efficiency of 1/3). In due course Reeves, Solberg and Smith very often obtained photographs of Jupiter at New Mexico where the resolution reached 1/2 of the theoretical resolution of their 300 mm telescope. Finally, it is at Catalina (1.5 m telescope) and at Pic du Midi (1.06 m telescope) that the highest resolutions have been attained: 0".2 or even 0".15 (but we are again below the theoretical resolution of the telescopes used).

In general, we can assume that the best lunar and planetary photographs show an average resolution of only 0".5, irrespective of the diameter of the telescope used (from 60 cm to 5 m!); and A. Dollfus has stated that the average resolution of planetary photographs is 0".6.

It is atmospheric turbulence that is responsible for the mediocre efficiency of large telescopes. Progress in photographic techniques has been phenomenally rapid in the amateur community. Over the years, however, the lunar and planetary photographs of amateurs have tended to be rather miserable affairs, with an average resolution from 2" to 4"! It was not until very recently, thanks to a miracle film, the best optics, the most rigid mountings, precision alignment and adaptive photographic techniques, that amateurs were able to make incredible progress, recording, with their telescopes of 20–40 cm, lunar and planetary details as small as 0".2, a resolution attained by the best professional observatories.

Solar photography at high resolution presents the greatest problem, as it is very difficult to combat daytime turbulence (seeing). Having placed their solar telescopes in particularly well chosen sites, the professionals frequently succeed in obtaining photographs showing details as small as 0".5, even 0".2. Amateurs are disadvantaged by their observing sites and thus by strong local turbulence; they are rarely able to exceed a resolving power of 0".8 (except for one indefatigable amateur astronomer who has already reached 0".3 and 0".25!).

High resolution photography has a fine future, especially as the use of CCD detectors in place of photographic film will improve results. High resolution is available to all who care to go to the trouble of mastering the techniques.

Part 1

1

Atmospheric turbulence

As a great optician once wrote, 'the worst part of a telescope is . . . the atmosphere'. It is a constant source of perturbation of the rays of light which reach us from the stars. According to Danjon and Couder (1935), the atmosphere exhibits defects of local inhomogeneity, due to winds, to local eddies and, above all, to inequalities in temperature either in the tube of the telescope or in the atmospheric strata themselves.

Because the refractive index of air varies with temperature, and the air temperature is inhomogeneous in the vicinity of, and above, the observing site, a ray of light passing through the Earth's atmosphere will be caused to deviate by a small unpredictable amount. The wavefront which reaches the objective is therefore more of an irregular surface than a plane, and the normal ray to each point on that surface will be in a state of constant fluctuation about a mean direction. The magnitude of the deflection of the normal ray can be taken as defining the degree of atmospheric turbulence (seeing). Meinel (1960) and Stock & Keller (1961) also explain the turbulence by the thermal heterogeneities of the atmosphere. According to its intensity, the turbulence entails an irregular motion of the image, or a considerable distortion of the Airy disk. Dollfus (1961) noted that the diffuse turbulent regions had a wavelength that varied between 10 cm and 25 cm. The influence of the air temperature is crucial: a layer of air 1 °C warmer and only 15 cm thick is able to shift the image in the line of sight by $\lambda/4$. With small telescopes (10–20 cm aperture) the image of a star can oscillate in the focal plane by between $0''.2$ and $3''$ (whereas, with large diameters the image becomes completely blurred). The heterogeneities in the refractive index of the atmosphere can be produced at various altitudes, up to 3000 m. The seeing varies with the wavelength of light and the zenith distance of the celestial object: short wavelengths are more perturbed and the images deteriorate as a function of zenith distance or as the atmospheric thickness increases.

According to Texereau (1984), there are various types of turbulence:

(1) *High-level turbulence*, which can be observed by placing the edge of the Moon in the telescopic focus and progressively defocusing the image (by increasing the separation of the eyepiece from the focal plane) up to the point at which we reach the turbulent layer (this phenomenon is perfectly visible on star images over a wide range of altitudes, sometimes up to 5000 m).

(2) *Local turbulence*, which is caused by the immediate environment of the telescope and which is due to diurnal heating by the Sun followed by nighttime radiation.

(3) *Instrumental turbulence*, which is produced even inside the telescope tube. We can demonstrate its presence by looking at a bright planet (Venus or Mars) and carrying out the Foucault test directly on the focal plane image: we search, by moving the head, for the place where the eye finds the point where the light beam is cut off by the edge of the shadow of the Foucault knife-edge. After several attempts, the slowly moving shadows, due to the warm air currents in the tube of the telescope, can be seen.

1.1 External turbulence

Danjon & Couder (1935) have proposed a good method for assessing external turbulence, which varies with the zenith distance. It involves observing, at high magnification, the diffraction pattern of a star. The appearance of the image, for a given aperture, allows for the turbulence to be defined in terms of the modifications of the Airy disk and its rings. We shall write t for the turbulence and a for the radius of the Airy disk. If t is small with respect to a, we obtain a diffraction pattern conforming with the theoretical one. As t increases, the theoretical diffraction pattern is progressively altered and allows us to define the quality of the images in a quantitative way. The turbulence will thus be estimated in fractions of arcseconds.

Assuming that we have a turbulence of $t = 0''.25$, we know that $a = 14/D$ ($''$). Knowing the

Table 1.1 *Appearance of the image as a function of turbulence*

		$t = 0''.25$
$t < 1/4\ a$	Perfect images, without visible distortion and little agitated	$D < 14$ cm
$t = 1/4\ a$	Complete rings, crossed by moving ripples	$D = 14$ cm
$t = 1/2\ a$	Medium turbulence, diffraction rings broken, central spot having undulating edges	$D = 28$ cm
$t = a$	Strong turbulence, rings weak or absent	$D = 56$ cm
$t > 3/2\ a$	Image tending towards a planetary appearance	$D > 84$ cm

diameter D of the objective, we can calculate the value of a, and thus the value of t, from the appearance of the star images, based upon Table 1.1 (Danjon & Couder). We therefore obtain the value of the turbulence in arcseconds. For example, if the images are perfect, without any deformation, t will be smaller than $1/4\ a$, so that $a = 1''$ for an objective of 14 cm. Thus, the turbulence is established to be less than $0''.25$, which indicates very good atmospheric conditions.

Following Danjon & Couder, if $t = 0''.1$ the images are excellent; they are good at $t = 0''.2$, average at $0''.3$, and poor at $0''.5$. It follows that we are only rarely able to use the theoretical resolving power of a 14 cm objective in an average location. In the greater part of the former French territories, t is generally greater than $0''.30$; its value does not fall below $0''.10$ for more than a few hours per year! The limit of resolution of a telescope of 10 cm will thus be reached very frequently, while that of a 80 cm telescope will hardly ever be reached.

The external turbulence is a function of a group of factors which directly control the image quality. The given climatic and geographical conditions allow us to predict the quality of telescopic images for a given site. We are certain to obtain bad seeing under the following conditions: large variations in temperature in 24 h, strong winds, continual and rapid changes in temperature, falling temperature throughout the whole of the night, a site placed upon a slope or in a hollow, etc. On the other hand, we can hope for good images when the temperatures are very stable, particularly at night, when the humidity is high and reaches the high atmospheric strata (fog is often favourable), and at the summit of an isolated mountain (if possible, surrounded by water). The ocean is often beneficial: on the borders of the sea with gentle breezes coming from the ocean, on small islands covered with luxurious vegetation and having a central elevation, etc. The 'haze' of the large cities sometimes plays a favourable role in improving the seeing (the images are often better in Paris than in the country). In many sites the images are best after midnight, when the effects of the Sun are completely nullified and when the air has become stable. One can often also get very good images at daybreak (Dobbins *et al.*, 1988). In reality there exist few really favourable sites with good images and the professionals spend years studying various sites before building an observatory. The amateur wanting to observe at high resolution must do as much!

There are, in fact, a certain number of sites which are well known for the quality of the high resolution photographs which have come from them: Pic du Midi (France), Catalina Mountain Observatory (Arizona), Pic du Teide (Tenerife) or Las Palmas (Canary Isles), New Mexico State University Observatory and Mauna Kea Observatory (Hawaii). The two observatories of Flagstaff are only in second place to these. Usually, the great observatories of Lick, Mount Wilson, Mount Palomar, Kitt Peak, McDonald, Yerkes, La Silla, Cerro Tololo, Mount Stromlo, Siding Spring, Haute Provence, Paris, Meudon, etc., which in principle were established at selected sites, are not found to be remarkable, except very occasionally, in the field of high resolution.

I was surprised to discover, in Equatorial Africa, sites characterised by great stability of the images, which seems to be linked to the climatic conditions found in a number of equatorial countries, where, unfortunately, there is not a single observatory.

But that is nothing new. In effect, as René Verseau told me, Bell (1922) wrote: 'The climate of Jamaica, albeit extremely damp, affords remarkably good seeing during a large part of the year, and permits the use of the telescope quite in the open, without inconvenience to the observer'. A little later, Boyer (1954), who was to discover the four-day rotation of the atmosphere of Venus, called attention to the good images that he was able to obtain at Cotonou (formerly Dahomey, now Benin) or even at Brazzaville (Congo). Later, Boyer (1958) published a comparative study of the turbulence at Pic du Midi, Cotonou, Brazzaville, Toulouse, Mont Louis and Auch (France). The hundreds of measurements were obtained by the method of Danjon and Couder, by

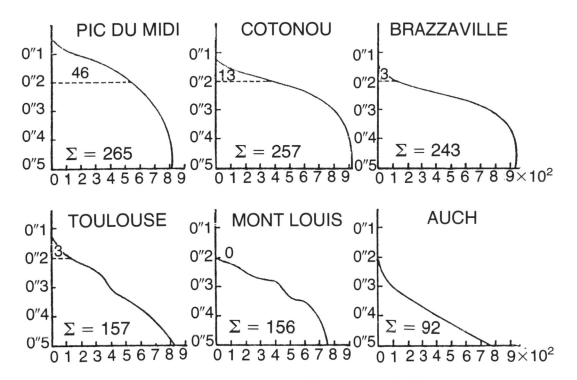

Figure 1.1. Graphical comparisons of atmospheric turbulence *measured at six different locations. Turbulence in arcseconds is plotted against the number of occasions when that degree of turbulence was* *found. The sum (Σ) gives the total value of the turbulence for that location. Note the high quality of the images at Pic du Midi, Cotonou and Brazzaville.*

Bartoli, Camichel and Boyer himself. The published graphs (Fig. 1.1) allow us to conclude that the Pic du Midi remains the best of the sites studied, because of the very low turbulence (between 0″ and 0″.20). Cotonou emerged on top for values of turbulence between 0″.20 and 0″.30, which is more interesting for the amateur (since it corresponds to good images with a telescope of about 20 cm).

If Brazzaville can give good images, the sea-level French stations exhibit turbulence exceeding 0″.40 for the greater part of the year. We can add to this the further advantage of the Pic du Midi with its low incidence of cloud and great atmospheric transparency. Against this, at Cotonou haze is present on every night and the transparency remains mediocre (we lose at least 1.5 magnitudes). On the other hand, the equatorial regions of Africa (Cotonou, Brazzaville, Makokou, etc.) enjoy a constant duration of their nights (12 h) and an ecliptic which passes through the zenith: that is to say, the Moon and planets are always high in the sky. R. Verseau was able to observe Mars over a long period with a 20 cm telescope, during the dry season at Brazzaville, and realised he had never seen the planet under such good conditions. In the region

of Leopoldville (opposite Brazzaville), the Belgian meteorologist Béruex (1958) studied the layered structure of the atmosphere during the dry season. He established the existence, in the temperate regions of the Congo (now Zaïre), of well-defined contraflowing air masses, a superposition of the numerous tropospheric depressions, of small thickness. The amplitude of the temperature variations remains low and the water vapour content of the air stabilises the temperature, at the various levels. The differential variations are sometimes only 1.5 °C, from the ground up to the 200 mb level. René Verseau specifies that in the dry season at Brazzaville, the winds are low (0–25 km/h from ground level up to 6000 m; studied by radiosounding). The atmosphere is so calm that the smoke from great bush fires rises vertically up to 6000 m above ground level.

Since 1963, I have been able to confirm the above facts, by observing Mars and Jupiter some distance from Makokou, in the middle of the Gabon forest, with the help of a 175 mm telescope: perfect images, no turbulence. Later, from 1968 to 1971, I was able to observe at Yaoundé (Cameroun) with a 250 mm telescope. As a result of the local geography (hills intercut by valleys) the images were clearly less good than those at ground level. All the same, it was at Yaoundé that I saw, for the only time in my life, the sky covered with stars without any scintillation! (The phenomenon lasted a good hour and the images, telescopically, were fantastic!) Later, at Cotonou (1977–84) I could observe (with a C8) the diffraction rings of Sirius on every night of the dry season. (My friend René Verseau was able to see them with me.) At Cotonou I could always see, with no difficulty, the companion of Sirius with a 400 mm telescope. With a C14 (355 mm), very often, the Moon and the planets exhibited no significant turbulence. We were able to photograph Jupiter, with success, on all clear nights, even in the rainy season.*

However, it would be wrong to think that it is like that everywhere in Africa. The worst images that I have ever obtained could be observed in the north of Cameroun, in the middle of the dry season. During the day, the ground and the dark rocks were heated to maximum temperature and released their heat during the cold nights (a fall of 35 °C in 24 h!). The images were so bad that no observation could be made; we could not even find the right focus of the telescope. Everything was completely blurred (with the turbulence exceeding 20″!). I lived for 4 years at Butare (Rwanda) in the hills, at 1750 m altitude. The sky was transparent and the nights very clear, but, despite the strong humidity at night (85%), the images were always bad, worse than in France, useless for high resolution photography (in spite of the Moon and planets being at the zenith!). This is due, in my opinion, to two major reasons: great changes in temperature in 24 h (up to 20 °C) and permanent winds, mixing up the layers of air of very different temperatures.

We can therefore conclude that the equatorial countries are much more variable than the temperate regions and that we can encounter both good and bad conditions. In a general way, one must search for humid regions, of constant temperature, covered by vegetation and free from winds.

Although less favourable, the tropical regions also often present excellent observational conditions for high resolution astronomy: 'In the latitudes 25–35°, of both hemispheres, the fraction of clear nights is the largest (with local exceptions) and the seeing is good to excellent a fair fraction of the time, particularly close to the western shores of various continents ... Tropical sites have a large fraction of good seeing on the nights that are clear ... At Bosscha Observatory in Java the seeing is almost always satisfactory with visual resolution of 0″.2 regularly achievable with the 60 cm visual telescope' (Kuiper, 1972). It is this that explains the proliferation of domes on the Pic du Teide or on La Palma (Canary Isles). It is no accident that the two best present-day planetary photographers observe from the humid, tropical regions.

Don Parker, at Coral Gables, Florida, enjoys an almost constant temperature, a strong humidity and the proximity of the sea; he always praises the quality of the images he obtains in his observatory and his photographs confirm this. Isao Miyazaki, living on the island of Okinawa, also benefits from a semitropical climate, hot and humid; he obtains excellent images. Dobbins *et al.* (1988) remark that the images are best when the sky is cloud-covered by day and does not clear before the beginning of the night (which happens very often in the equatorial regions, during the dry season). The tropical zones Florida and Okinawa have the disadvantage of occasional periods of very unsettled weather, with violent cyclones. The altitude of the ecliptic is less favourable than in the equatorial regions (which remain, for me, clearly superior). In my opinion researches into the seeing conditions should be made in the great Gabon forest, as well as on the coast (Libreville) and on the mountains of the interior (Belinga, Diamond Mountains, etc.).

Atmospheric turbulence is much more important by day. Solar photography, at high resolution, therefore remains extremely difficult. Cortesi (1974), at Locarno-Monti (Switzerland), was able to make 4600 estimations of the daily turbulence between 1950 and 1973. The daily turbulence varies between very wide limits: from 1″ to 40″! At Locarno-Monti the best images were found in summer (July–August), in inverse correlation with the zenith distance. The stability seemed optimal 2 h after sunrise. A resolution close to 1″ could sometimes be obtained, after several days of high pressure, with almost no wind. This is also my finding, in my present observing site, in the Montpellier region.

It is interesting that Cortesi considers that the images have progressively deteriorated since 1967. The same remark has been made by other observers, notably by Viscardy (1987), who thinks

---

* I spent 57 days at Cotonou in January–February 1993 in order to take high resolution photographs. I had the disappointment of establishing that the climate had changed considerably (owing to the advance of the desert, deforestation, etc.) and the seeing was not nearly as good as in 1954 or even in 1977–84. I no longer recommend Cotonou as a special site for high resolution photography: of 5200 negatives obtained in 1993, only about 40 were worth keeping.

that atmospheric turbulence has progressively increased during the last 30 years. I have come to the same conclusion about Cotonou (see footnote, page 6).

Bray & Loughead (1964) considered that during the daytime the exterior turbulence augmented the local turbulence (thermal convection, from the heated ground) and instrumental turbulence (direct heating of the telescope by the Sun). The Australian astronomers remark that, although the daytime turbulence varies between about 1″ and 5″, there are often fleeting moments of calm. They obtained the best images close to the meridian (and not 2 h after sunrise, which is the opinion of the majority of observers), as they have succeeded in overcoming local and instrumental turbulence (see Section 4.2.2). In fact the telescope looks across a veritable 'forest' of disturbed, convective columns. At midday, when the Sun is at the zenith, we can sometimes pass between the columns, which are then parallel to the telescope tube, and so benefit from a less disturbed region.

The climatological and geographical conditions exert a considerable effect on the magnitude of the daytime turbulence. Surroundings covered with dense vegetation, an observatory raised above the ground (like a solar tower) and a site with small temperature variations all constitute favourable factors for obtaining good images. The ideal is a small oceanic island, densely covered with vegetation and having a high, isolated mountain (e.g., Canary Isles, Hawaii). If the high mountain includes a crater lake at its summit, the observatory could be installed in the centre of the area of water. As in the case of night-time turbulence, persistent high pressure and high humidity would be the most favourable factors for obtaining the best images.

1.2 Local turbulence

Local turbulence also plays an adverse part which one must try to eliminate. By night, as by day, we must do everything to avoid the heating of the telescope, which may often be very difficult to achieve. A plastic tarpaulin or similar protective sheet is the worst covering for a telescope, as it gets very hot in sunlight, and also collects night-time dew, which condenses on the instrument. Under a tarpaulin in midsummer in central France, the telescope tube can reach 45 °C or more. One has to wait hours before being able to observe, and it is often not until after midnight that a state of thermal equilibrium is attained (or only in the small hours, if the temperature falls all

night long, which is the case at many sites). We can reduce the solar heating of the covering, and thus of the telescope, by wrapping it with sheets of aluminised plastic (Mylar film, as used in 'survival blankets', etc.). When one cannot avoid using a tarpaulin cover, it is preferable to leave the cover off during the day in dry weather (at least if the telescope has not been painted black). The most generally adopted solution is to use a wooden mobile shed. In Equatorial Africa, when the daytime heating was perhaps harmful, I used lightweight wooden sheds on wheels. At Cotonou they were simply painted white and provided with a double roof. I was able to study, over a 24 h period, the change in temperature inside the hut, as well as that of the main mirror of the telescope. As can be seen from the graph in Fig. 1.2, the telescope equilibrated very quickly, after the sun disappeared. However, the climate of Cotonou is very special: the variations in temperature remain very small, notably during the night.

On the high hills of Rwanda the situation was quite different: great variations in temperature (up to 17 °C in 24 h) and a steady fall in temperature through the night. Thus, one must consider providing much better thermal protection for the telescope. The hut which I had made had seven large windows constantly open and an internal double wall: wood painted white outside and aluminised plastic inside (with a 5 cm air-gap between them). In these conditions the temperature inside the shed never exceeded 24 °C (Fig. 1.3). Invariably the telescope remained unusable for the first few hours of the night. It would be necessary to install air-conditioning in the shed! Today a new type of paint, obtainable in the USA, the so-called 'IR (infrared)-blocking paint', is able to considerably reduce the solar heating of a shed or even of a telescope (Beish, 1992). The dream of many amateurs is to have domes for their telescopes. Domes have a strongly aesthetic appeal, but, in my opinion, are very much inferior to the mobile shed. Even if we protect them as much as possible (aluminium dome, painted with titanium oxide paint or with the new IR-blocking paint) the solar heating and the night-time chill will cause problems across the slit or opening: we must not forget the warm air currents, escaping through the slit around the opening of the telescope! A dome is poorly suited to high resolution work, and the problems of local turbulence will always be ineffectively dealt with. Observatories with opening or sliding roofs are sure to have better conditions and protect the instrument from wind. The local turbulence

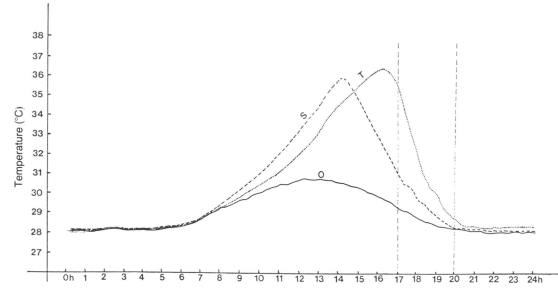

Figure 1.2. The diurnal temperature variation *at Cotonou (Benin): O = external air temperature; S = temperature inside the observing hut; T = temperature inside the telescope tube. It can be seen that the* telescope cools down very quickly and that observations can take place from 20 h in the evening. During the night, the temperature is constant, leading to excellent seeing at this station.

Figure 1.3. Mobile shed for a telescope, *guaranteeing a high level of protection against the daytime heat (in Butare, Rwanda), by means of a double roof (r), six windows (w) and an interior* insulation obtained with a sheet of Mylar film (a), placed several cm from the wooden lining of the shed: 1, exterior; 2, interior.

perhaps is also caused by 'animal heat', given out by several observers crowded into a hut or a tiny dome (in the same way one must avoid having any non-observer standing beneath the telescope tube).*

1.3 Instrumental turbulence

Instrumental turbulence depends essentially on the local turbulence. If the telescope is in thermal equilibrium with the surrounding air, one cannot have serious instrumental turbulence. For several years, numerous commercially made telescopes (Celestron, Meade, etc.) were painted black, which surprised many observers; traditionally, telescope tubes in general were always painted white or pale grey. In fact, as Beish (1992) remarks, a telescope painted black will cool down very quickly at night (but it must not be exposed to the sunlight during the day). On reflection, I do not favour this innovation, as a telescope may also be in use during the day; by night, one is more likely to bump into a telescope painted black! It is, besides, as unaesthetic as possible.

If the telescope is of small size and is kept (during the day and in winter) in a heated house, one must wait several hours for it to cool to ambient temperature after taking it outside, and longer still if the telescope has a closed tube: Schmidt–Cassegrain, Maksutov or refractor. It is useful to place a precision thermometer (a mercury thermometer graduated in tenths of a degree, as used in developing colour films) inside the telescope tube, in the vicinity of the main mirror. As long as the temperature, inside the tube, remains 3–4 °C above that of the air outside, the images will be very poor. If the telescope has an 'inspection door', one can speed up the cooling of the main mirror (which is the last thing to cool down, by virtue of its high heat capacity) with the aid of an electric fan. At Pic du Midi, with the 106 cm telescope, small electric fans are in use nearly all the time, even during observation. Finally, manufacturers have tried to improve the thermal equilibration of their telescopes by forced ventilation (225 mm Schmidt–Cassegrain, by Takahashi) or by using a system of openings behind the main mirror (Mewlon Cassegrain, by Takahashi). The large refractors are often very

difficult to cool down, as their modern cemented triplet objectives have a great thickness of glass at their centres. The Astro-Physics, Zeiss, Vixen and Takahashi triplets are 'cemented', not with Canada Balsam but with oil of suitable refractive index to avoid reflections, improve the light transmission and simplify the correction of local polishing irregularities. At sites where the temperature falls rapidly throughout the night, a heavy mirror or an objective never completely reaches thermal equilibrium. We probably cannot obtain the best possible images from them. It is therefore important to understand the night-time changes in temperature of the observing site. Often, the temperature stabilises during the second half of the night. In summer it is practically impossible to obtain good images at the beginning of the night: the ground, the covering and the telescope itself are all warmer than the surrounding air. The majority of the very best lunar photographs are obtained during the last quarter Moon, in the final hours of the night. In humid regions dew can be deposited on the corrector plate of the Schmidt–Cassegrain, the objective of the refractor or the secondary mirror of a reflector. One must especially avoid using an electric hairdryer to remove it, or additional instrumental turbulence will be guaranteed! The only defence against dew is a long plastic dew-cap, with a double interior lining of absorbent black paper. Tuthill makes this item in the USA, which also allows the use of a low-power resistance heater inside the dew-cap, but it is desirable not to use the latter accessory, especially for high resolution photography.

To conclude, the work of the astrophotographer bristles with often nearly insurmountable problems: he or she must find a site with low turbulence, and then eliminate all the sources of local or instrumental heating.

1.4 The choice of an observing site

Professional astronomers today spend years studying several apparently favourable observing sites before building their observatory. The number of these sites is limited, in proportion to the increasing requirements of the astronomers. So we witness today a concentration of giant telescopes in just a few privileged places: Hawaii, the Cordilleras (Andes), the Canary Isles, etc. In the first place, professional astronomers look for the greatest possible number of clear nights per year: anything from 280 upwards, preferably, more than 320! That is why the largest observing

---

* It was this 'dome effect' which once caused Barnard to remark that he wished the dome of the great Yerkes refractor could be mounted on a rail, so that he could push it away and work in the open air with perfect images! Such was the problem of local turbulence in his view. (RJM)

9

centres are often found in desert regions with a very dry climate. As they have good weather more or less permanently, the desert regions ensure that the atmosphere will be very transparent (owing to the absence of water vapour), and this advantage is enhanced by high altitude: 2000–4200 m. The majority of modern astronomical research necessitates a large number of clear and transparent nights, with a site preferably not too far from the equator in order to cover a good part of the sky, everything benefiting from nights of similar duration throughout the year. The problem of turbulence is therefore not the prime consideration. In stellar astronomy a resolution of 1″ is often considered to be good enough (more especially as the giant telescopes are, in general, not capable of realising their theoretical resolution).

In high resolution astrophotography the image quality is of fundamental importance. Unfortunately, the great observatories are only rarely able to benefit from nights of very low turbulence (less than 0″.20). Even at Pic du Midi, we can obtain very poor images, and Hawaii does not always fulfil its promises. We know of very few high resolution planetary photographs obtained in the Andes. From personal experience (and a critical examination of the works of Slipher), Flagstaff is *not* a high resolution station according to modern standards. I was very surprised to learn that there is no observatory, for high resolution, in the humid, equatorial regions (which, unfortunately, are disadvantaged by poor transparency and a limited number of useful nights).

In our times, amateurs wishing to become high resolution astrophotographers will have an interest in undertaking a preliminary survey before installing an observatory. Few of them have the opportunity. To my knowledge, only G. Viscardy chose his site specially at Saint-Martin-de-Peille (above Monte Carlo) before installing his telescopes. His choice proved to be a good one, even if the quality of the images was not as high as our colleague had wished. D. Parker and I. Miyazaki have found themselves by chance in favourable sites, and their photographs demonstrate the quality of the images they enjoy.

In the majority of cases, however, the amateur astrophotographer is obliged to content himself with the observing conditions imposed by his place of residence. Some of them are only able, sometimes, to take a transportable instrument to a better site. Certain sites are so bad that all serious work is impossible. (This was my situation when I was living in Rwanda. The same applies now at Saint-Clément, Herault, in the south of France: I was never able to obtain a real high resolution photo, in 2 years of assiduous practice!) However, at many apparently quite mediocre sites favourable conditions may be enjoyed on a few nights per year (misty nights, long periods of high pressure, etc.). In such cases suitable conditions must not be wasted and nothing must go wrong! From this it follows that the painful duty of those who want to obtain high resolution photographs must be never to waste a single clear night and always to be ready to take advantage of fortuitous moments of good seeing. From this point of view, we can divide observers into two categories: those who insist that, where they live, nothing good can be done because of constant poor conditions (this is the case of the vast majority of people, who only think about observing from time to time); and those (very few in number), often living in the same places as those in the first category, who stubbornly obtain, three or four times a year, high resolution photographs.

The Sun, Moon and planets are very brilliant objects that do not require great transparency for their observation. That is often why large cities are favourable for obtaining high resolution images, as the haze with which they are overlaid (often up to high altitude) has the advantage of stabilising the atmospheric layers, especially at certain times during the year. We often obtain evidence that the Paris region, for example, is not unfavourable for planetary observation. Furthermore, it was at Meudon that the 'Grande Lunette' of 83 cm aperture allowed the famous E. M. Antoniadi to observe Mars in the most remarkable detail. It was also in the city of Paris that several of the best lunar photographs were taken, by C. Arsidi and G. Thérin. Finally, the famous German amateur G. Nemec obtained, almost 30 years ago, excellent lunar and planetary photographs from the middle of the city of Munich.

I hope I will not be accused of talking nonsense if I insist here that one should be able to enjoy a few hours of good images almost anywhere, if one can take the trouble to be at the eyepiece every clear night throughout the year. High resolution is the prize! That is why there are very few good Solar System photographers.

Must one own a proper observatory? For the person who wishes to obtain high resolution photographs, as certainly and as frequently as possible, it is unimaginable to think of mounting and dismounting the telescope at each observing session. Thus it is obviously preferable to have access to a garden, a terrace or a large well-placed balcony, so that the telescope can be installed in a permanent manner.

Many amateurs imagine that it is essential to have a dome to do serious astronomical work. Following Kuiper (1972), we can improve the image quality by constructing a relatively small dome, opening at the top and bottom and covered with a good coat of titanium oxide white paint. Four or five powerful ventilators will keep the air inside the dome well mixed. Some of the best high resolution astrophotographers neither have, nor wish for, domes: a mobile shed is perfectly adequate and, in combating local and instrumental turbulence, much to be preferred. Revolving huts, such as that of D. Parker or my various mobile sheds, are very convenient. We know of many superb domes which have never produced anything of value!

We are therefore not being 'snobs'. We have to protect our instrument from the excesses of daytime heating. Depending upon the amount of money we have available, we can install different types of shed, from the simple wooden hut on wheels (like those I had made in Africa: Fig. 1.3),

very light and costing little (but useless in areas prone to cyclones!), up to the brick-built observatories provided with an opening dome, mounted on rails.

This last solution (to make use of a lightweight shed) is the best, as the telescope is partially protected from the wind but remains under an 'open sky', because of which thermal equilibration is rapidly established, with no problems due to the 'dome effect' during observation. When the site is continuously swept by a light but constant wind (as at Cotonou, where the breeze, coming from the south, blows directly upon the telescope), we can provide some mobile wind-breaks. A fixed shelter also allows us to work in comfort, perhaps to make use of cupboards, a small table and (in my opinion most important), an observing chair of variable height (Fig. 2.2:3). It is not always necessary to have mains electricity supplied, the majority of modern telescopes being controlled by stepping motors, which work off a 12 V supply. However, if using CCD cameras and computer equipment is envisaged, then a mains supply will be necessary. Those who wish to have precise constructional details for an observing shed can find them in Krick (1992).

2

The telescope

2.1 Diffraction patterns, theoretical resolving powers, etc.

All astronomical telescope objectives give, when in focus, the diffraction pattern of a star (Danjon & Couder, 1935; Sidgwick, 1980; Texereau, 1984; etc.). Instead of an infinitesimal point, which must represent the stellar image, we can see a false disk, or Airy disk, whose angular radius is given by the formula $a = 1.22\ \lambda/D \times 206.265$ (in arcseconds). The angular radius of the Airy disk is therefore inversely proportional to the diameter of the objective. For white light we can take λ as being fixed ($\lambda = 550$ nm); therefore, $a = 14/D$ (where D = diameter of the lens or mirror in cm). The diffraction disk given by a 14 cm telescope measures 2 arcsec across.

The diffraction disk is surrounded by one or more diffraction rings (Fig. 2.1). The total light energy is divided up as follows: Airy disk, 84%; 1st ring, 7%; 2nd ring, 3%; etc.).

When the telescope has a central obstruction due to the presence of a small secondary mirror (Newtonians, Cassegrains, Schmidt–Cassegrains, Maksutovs, etc.), the diffraction pattern is altered to a greater or lesser extent; the diameter of the Airy disk decreases but the rings become brighter.

Perfect Optics

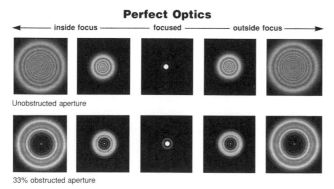

Figure 2.1. Diffraction images, *showing the Airy disk and its rings. (Focused image enlarged for clarity.) A central obstruction, such as the Newtonian secondary, increases the complexity of the out-of-focus diffraction images.* (Reproduced with permission from *Star Testing Astronomical Telescopes: A Manual for Optical Evaluation and Adjustment*, by H. R. Suiter, Copyright © 1994 by Willmann-Bell, Inc.

Table 2.1 *Distribution of relative light intensity in the diffraction pattern (Rutten and van Venrooij, 1988)*

Central obstruction (as percentage or fraction of diameter of objective)	0% (0.0)	25% (0.25)	50% (0.50)
Brightness of the Airy disk	84%	73%	48%
Brightness of the first ring	7%	18%	35%
Brightness of the other rings	9%	9%	17%

Table 2.2 *Theoretical resolving powers (after Dawes)*

Diameter of objective (cm)	6	12	18	24	36	50	100	200
Resolving power (arcsec)	2	1	0.60	0.50	0.30	0.20	0.10	0.05

Table 2.1 shows that a large central obstruction considerably alters the diffraction pattern: with 50% obstruction (never realised in practice) there is the disadvantage that there is more light energy in the rings than in the Airy disk. Under such conditions, the resolving power is altered, as we shall see later. Knowing the size of the Airy disk allows us to define a theoretical resolving power, which is the quantity which allows us to see the true nature of a double star, if the angular separation between the centres of the two components is greater than 0.85 of the radius of the Airy disk. The theoretical resolving power (Table 2.2) thus has a smaller value than the radius of the diffraction disk. The formula of Lord Rayleigh allows us to calculate the theoretical resolving power (or separating power) of an objective of known diameter: $R = 13''.84/D$ (cm), which gives a resolving power of $1''$ for a 14 cm objective. Dawes, after observing many double stars, proposed the following empirical formula: $R' = 11''.58/D$ (that is to say that an objective of only 12 cm aperture will separate double stars with components $1''$ apart). The Dawes formula (known as Dawes' limit) is the most often used.

We can therefore see that for a given wavelength (white light is taken to have $\lambda = 550$ nm), the theoretical resolving power is inversely proportional to the diameter of the objective. However, the resolving power also varies with the wavelength. Thus, for violet light ($\lambda = 360$ nm) the formula becomes $9''.07/D$, while for red light ($\lambda = 770$ nm) it is $19''.4/D$ (Sidgwick, 1980).

As we have seen above (p. 12), the central obstruction can exert a favourable influence upon the ability of an instrument to separate fainter double stars (thanks to the reduction in the diameter of the Airy disk). In contrast, on a bright planetary feature, the separating power is diminished, since an important part of the light energy is transferred to the rings. For a central obstruction of 50%, it is the diameter of the first ring that determines the ability to separate feebly contrasted details, particularly if – in photography – the image is overexposed.

For moderate central obstructions, we can readily establish a reduction in the image contrast. According to L. T. Johnson (in Dobbins *et al.*, 1988), this reduction will be very noticeable: for an obstruction of only 25% (0.25), the loss of contrast is considered to be 50%. This seems to me strongly exaggerated. Everybody can make the following experiment: observe a planet, in good seeing, with the aid of a refractor between 10 cm and 20 cm in diameter having available, at the centre of the objective, small disks of black paper, with diameters of 0.20, 0.30, 0.35 (etc.) that of the objective. It will be easily seen that, for an obstruction of 0.20, the human eye will not readily notice any difference; a loss of contrast only becomes appreciable with an obstruction of 0.35 (moreover, this loss of contrast is perhaps easily compensated for in astrophotography). Rutten and van Venrooij (1988) think that an obstruction of 25% (0.25), which is the most usual in Newtonians and Cassegrains, entails a loss of contrast of only 15%. On the other hand, an obstruction of 50% makes for a contrast loss of at least 50%, reducing the resolving power on extended objects. Obstructions of between 0.30 and 0.34, very common in the short Cassegrains, Schmidt–Cassegrains and some Maksutovs, involve a loss of optical quality equivalent to an accuracy of $\lambda/4$ of the whole optical system. This amounts to saying that a telescope made to a precision of $\lambda/12$ automatically becomes only $\lambda/4$ if it is obstructed by 32%; if the optics are only accurate to $\lambda/4$ at the start, the detrimental influence of the obstruction will thus be much more noticeable. But all this is theory, and the opinions of those who use telescopes are more varied: A. Kutter (the inventor of the telescope with folded optics) did not allow the central obstruction to exceed 10%, whereas H. Dall accepted 20% and Steel and Brouwers between 25 and 30%! However, almost all authorities agree that, for a strongly obstructed optical system, the rule of Lord Rayleigh (see earlier) no longer applies and that the optical system would be quasi-perfect (which never happens in reality). In my opinion the problem is very complex and we shall come back to it several times. We must not forget the 'practical' aspect, as we know many astrophotographers who obtain, with telescopes of central obstructions of between 0.30 and 0.34, truly extraordinary high resolution photographs. Thus, we distrust any hasty conclusions in this field!

We shall see that it is the diameter of the telescope objective that uniquely determines the angular resolving power of a given instrument (performance being diminished, again in a way very poorly known, when the objective is affected by a central obstruction). We have, up till now, only considered the theoretical side of the question, supposing that the objective is optically perfect. Figuring a perfect mirror or lens is extremely difficult. Following Lord Rayleigh, for an astronomical objective to give satisfactory images, it must be accurate to within a quarter-wavelength of its theoretical figure (whether it be spherical, parabolical, elliptical or hyperbolical). Modern observers always hold that the limit established by Lord Rayleigh is insufficient, when it is a question of observing low contrast objects (planetary surfaces, for example). High resolution requires a precision of $\lambda/20$; some consider that at $\lambda/4$ the contrast falls to 50%, which seems unlikely to me. It is possible that, in precise instances, these demands can be shown to be justified (and they are especially so when it comes to space telescopes, which are not affected by atmospheric turbulence). However, it is true to say that there are very few amateur instruments, and even fewer large telescopes, whose overall optical quality is better than $\lambda/10$. It is not too difficult to successfully figure a 200–400 mm mirror with a true precision of $\lambda/20$. But we must make the other optical components equally perfect, and this is rarely the case, especially when the telescope is a complex one. If we desire quasi-perfect optics for high resolution photography, with excellent images, the ideal remains the simple Newtonian, equipped with a mirror of $\lambda/20$ (or better) and a quasi-perfect plane mirror. This system is perfectly

Table 2.3 *Theoretical resolving power in the focal plane of an objective as a function of focal ratio*

Focal ratio $F/D = f$	2	2.8	4	5.6	8	11	16	22	32	45	90	180
Resolving power (in fractions of mm (lines per mm))		750	530	370	270	190	140	95	70	47	33	17	8.5

attainable in practice. On the other hand, it is not always so in the case of a refractor with a triplet objective or for a Schmidt–Cassegrain having three optical components, one of which is an aspherical window.

The resolving power can perhaps also be calculated, not in angular measure, but in fractions of a millimetre in the focal plane. The formula now becomes $R = 1.22 \times \lambda \times F/D$, where F is the focal length and D the diameter, which can also be written as $R = 1.22 \times \lambda \times f$, where f is the focal ratio, F/D. Thus, in white light ($\lambda = 550$ nm) an objective with a focal ratio of 16 yields a separating power of 10 μm, i.e. 0.01 mm or 1/100 mm: this could represent the capability either of an enormous telescope or of a tiny 8 mm ciné-camera!

Table 2.3 is instructive. It shows us why very fast Schmidt telescopes, of low f number, show much superior resolving power than the best photographic films; why, at the focus of a Newtonian working at $F/D = 5$, Kodak TP 2415 film is indispensable if one wishes to make use of the resolving power of the main mirror; and also why, at $F/D = 90$ (often used in planetary photography), we are able to use almost any film, without worrying about loss of resolution.

Above all, we must not forget that, whether we use a photographic objective or a telescope to record an image on a photographic film, the final resolving power becomes $1/R = 1/R$ (obj.) $+ 1/R$ (film) i.e. it will be found to be always lower, even if the film is a high performance one.

2.2 The relation between the theoretical resolving power and the visibility of fine, isolated details (definition)

The formula for the theoretical resolving power tells us something precise: with a telescope of given diameter, we are able to recognise the double nature of two very close stars (and the magnitude of their separation). However, the resolving power does not tell us precisely about the chance of seeing (or photographing) an isolated detail (see W. H. Pickering, 1895, 1920, cited by Sidgwick, 1980; also Dragesco, 1969, 1992). Some are surprised to learn that we are

Table 2.4 *Visibility of details of various form and contrast: resolving power of the naked eye (limit of visibility)*

Object observed	Maximum contrast	Contrast 0.75	Contrast 0.45
Black disk on bright background	31″	35″	43″
White disk on black background	25″	28″	34″
Black line, 100 times longer than broad (on bright background)	5″.5	7″	10″

able to see and photograph details below the theoretical resolving power. In this field, a very specialised one, precise experiments have been made – some with the naked eye, some telescopically – for more than a century. It is well known that a young eye has an average resolving power of 60″, from the results of periodic Foucault tests, at maximum contrast. I undertook in 1942 the experiments summarised in Table 2.4 (the separating power of my right eye being 56″, in white light).

Examination of Table 2.4 shows that we can see very much smaller details than the theoretical resolving power. We can easily see bright objects on a dark background and especially very elongated objects. We also consider that we can see a black line, on a white background, when its width is just 1/10 of the resolving power. This is nothing new, for Maunder & Maunder (Sidgwick, 1980) were able to see with the naked eye a telegraph wire against the sky when its diameter fell below 1″. This experiment was repeated by Fabry & Arnulf (1937), who saw perfectly, with the naked eye, a single telegraph wire against a bright sky, when it subtended only 1″ (that is to say, 60 times better than the theoretical resolving power of the eye!). Identical results can be obtained by using astronomical instruments: with the 28 cm Harvard refractor it was possible to see a human hair, on a bright background, when it subtended only 0″.29.

These results are extremes, never surpassed. We know, however, that the Cassini Division in Saturn's rings, 0″.5 wide, is visible in a 60 mm telescope, whose resolving power is only 2″. A 150 mm telescope, whose resolving power is no more than 0″.8, allows a lunar rille as narrow as 0″.2 to be seen. Dobbins *et al.* (1988) note that the famous Rima Tenuis on Mars (a rift in the north polar cap) was observed several times with 20 cm telescopes, when it measured scarcely 0″.1 across. I shall add (and I shall return to this in Section 4.3.5) that the rille in the Alpine Valley, on the Moon, which is no wider than 0″.27 on average, can be seen with a 20 cm telescope (resolving power 0″.6).

On the subject of resolution (or, rather, of definition) of a lunar photograph, I must add that it is a relative matter: we measure the finest detail visible on the image (the width of a long, well-contrasted rille, or Rima, for example), but the resolution found refers solely to this detail. I have already written that the resolving power of my eye, under the Foucault test (using parallel black and white lines, at maximum contrast) varies from 56″ to 63″. But if I look at a landscape, under a cloudy but well-lit sky, I am able to see a telegraph wire, in the distance, even if its width does not exceed 2″ (thus 30 times better than the standard resolving power). On the other hand, the same day, in a darkened wood, I am unable to differentiate weakly contrasting details, subtending less than 3 arcmin (or 180″). Thus, my eye is able to yield a resolving power between 2″ and 180″! It is so even for astrophotography. We can define a multitude of resolving powers on the same photo: a maximum resolution of 0″.25, on a long fissure close to the terminator, and only 1″.5 on the less contrasty domes, far from the terminator.

2.3 Testing the optical quality of a telescope

Be that as it may, it is very important to know the true quality of the instrument which we intend to use for high resolution astrophotography. For us telescope users, it is not a question of dismantling the instrument in order to test each separate component. This is beyond the capabilities of most amateurs (except with Newtonians, a study of whose mirrors can be made by the user; an extensive literature awaits them – see Texereau, 1984). All observers can get an idea of the overall quality of their telescopes by observing the sky.

In the first place, the diffraction pattern of a star must be observed. This examination must be repeated a large number of times, as the result depends on factors other than the quality of the

telescope: optics not reaching thermal equilibrium, strong turbulence, presence of wind, etc. The ideal is to look at a 4th magnitude (or fainter) star, close to the zenith, under good seeing. It is necessary to use a high power (twice the aperture in mm: ×400 for a 200 mm mirror) and, if possible, an excellent eyepiece (as eyepieces can often be awful), such as a Plössl, Orthoscopic or Nägler. Avoid using a star diagonal. Observing directly, in the region of the zenith, is uncomfortable with a refractor or a Cassegrain, but it must be done! The star, placed in focus at the centre of the field of view, should resemble the textbook appearance: a tiny Airy disk, absolutely circular and of uniform brightness and with at least one absolutely circular diffraction ring of uniform brightness (apart from the undulations due to bad seeing). Atmospheric turbulence rarely allows us to see a stable image of the Airy disk and its rings (particularly if the telescope is larger than 20 cm in aperture). Thus, one must look frequently, at different times of the night. Therefore, one must have plenty of time available to judge impartially a given telescope. Eventually, on a good evening, we shall be agreeably surprised to see a textbook diffraction pattern. That will be the first clue concerning the quality of the telescope. In contrast, if the Airy disk is oval, if the first ring is non-circular and irregular, and particularly if bright patches are superimposed upon the Airy disk or rings, or if a luminous haze surrounds the image, we know that the instrument is defective. It is more important to observe, attentively, the extrafocal and intrafocal images: they must show the same appearances when the image is progressively defocused, either towards or away from the main mirror. This observation can be conclusive, even with mediocre images. The least spherical aberration can be seen at once: the extra- and intrafocal images will be different (Fig. 2.1). As a result of the observed aspects, we know whether it is a question of undercorrection or overcorrection of the spherical aberration.

As already stated, we must repeat this observation a very large number of times, at various hours of the night, because, if the objective is very warm (and this is especially true for a cemented triplet objective) in comparison with the surrounding air, it can give images which could be thought to be due to spherical aberration. In the same way, diagonal reflections with prisms can introduce a small spherical aberration. It needs a good month of observation to give an exact idea of the quality of the diffraction images given by the instrument under test. If the images are perfect, i.e. conforming to theory, in the focal plane as well as extrafocally, we can conclude that the telescope is sound, and probably 'diffraction limited'.

If one considers, after several attempts, that there exists a small spherical aberration (intra- and extrafocal images slightly dissimilar) but if the image at focus seems correct, the telescope can be considered to be usable, although imperfect. On the other hand, if the spherical aberration is strong and if the focal plane image is bad, even under very good seeing (difficult to realise if the telescope exceeds 300 mm in diameter), such as a blurred spot surrounded by bright areas of various sizes, no diffraction rings, etc., the telescope must be rejected. We must add that a telescope of 800 mm, used at an ordinary site, will never give a textbook diffraction pattern, irrespective of its quality, the night-time turbulence literally destroying the classic Airy disk. On an even larger telescope, we are able to base our conclusions only on a study of the extrafocal images, without ever having the certainty of knowing the instrument well. Even a 400 mm can demand months of observation before we have the power to make a good estimate of the optical quality.

A second test concerns the focusing of a brilliant planet. (The planet Venus near superior conjunction or Mars, when it is far from opposition, will be suitable; the test requires a tiny bright, round image.) The procedure is simple in theory but difficult in practice: it will suffice to cut the pencil of light rays, in the focal plane, with the aid of a thin metallic blade (or knife-edge), placing the blade across the opening of the drawtube. By delicately adjusting the position of focus, it is possible to find the exact place where the objective, uniformly illuminated, is 'extinguished' all at once. Just before this point, one can see, for two or three seconds, the faults of the optical system used: instead of a uniform disk, one will observe either a gentle parabola or shadowed regions, dimly perceptible circular rings, internal defects in the glass (Schmidt–Cassegrain windows), etc. In reality, however, the turbulence is always too strong and tends to mask the appearances due to small optical defects, for which we are trying to obtain evidence. Therefore, this test is extremely difficult, as it requires perfect images, an excellent drive and considerable experience in the interpretation of Foucault images. When I tested several Schmidt–Cassegrains (Dragesco, 1978) by this method, I had to work in an 80-m-long tunnel, with the aid

of an artificial star. It is always possible to eliminate the effects of atmospheric turbulence by adopting the method used by Jean Texereau, when he was checking and refiguring the optics of the telescope of the McDonald Observatory. He had to make a small mounting, allowing for photography of the image under the Foucault test, of a bright planet or Sirius. A long exposure (deliberately chosen: 20–30 s) cancelled out the effects of rapid turbulence. The mount is not hard to make: a Foucault knife-edge and a micrometer screw are installed in the drawtube and the image is captured by a 100 mm or 150 mm objective, with a standard reflex camera body. The big difficulty is in having to keep the image of the planet on the knife-edge throughout the time exposure. I know of no amateur who has perfected this type of Foucaultgram. There remains the possibility of working in a laboratory to obtain a Foucaultgram, by means of an artificial star, placed more than 100 m away. We can get rid of local turbulence by means of long exposures (up to 2–3 min). Alternatively, we can attempt the autocollimation test, with the aid of a perfect plane mirror (very hard to find!), as di Cicco (1989) has done.

Be that as it may, before contemplating high resolution astrophotography, we must be sure we have an instrument of acceptable quality. Those who do not like to put their own telescopes together can draw upon the expertise of qualified opticians.

In the field of high resolution, and in spite of atmospheric turbulence, the use of high quality optics is absolutely essential. That is more important than seeking to possess a very large, but poor quality, instrument. Today we can be certain that an excellent 30 cm telescope will allow a skilled worker to obtain high resolution photographs, which will stand comparison with some of those taken by the world's most famous telescopes.

2.4 What aperture is best for high resolution photography?

Following the definition given on p. xiii, high resolution in a relative sense can be obtained irrespective of the diameter of the instrument, the goal being to photograph the smallest details accessible to the telescope in question. Thus, high resolution can be envisaged even with a small 60 mm refractor. However, if we desire to photograph the smallest details accessible to the greatest instruments (severely handicapped by turbulence), we must try to define an optimum diameter.

For solar photography, it is useless to seek to use a really large instrument. The daytime turbulence is always very strong. Few amateurs are able to do better than a photographic resolving power of 1″: that is to say, a refractor of 12–13 cm aperture suffices 90% of the time. It was with a 12.5 cm that Bray & Loughead took thousands of high resolution solar photographs in Australia. Exceptionally, Rouvière (1979) was successfully able to use a 20 cm solar telescope (in a selected site), whereas W. Lille very regularly obtains magnificent solar photographs with very high resolution with simple lenses of 175 mm and 300 mm aperture! However, the German amateur from Stade is probably a unique case. In especially well chosen sites we can hope for a few days per year when we can use the full resolving power of a 300 mm telescope, but the standard solar instrument must be a refractor between 130 mm and 180 mm in diameter.

For lunar photography, which provides us with relatively contrasty details, it is not necessary to use very large telescopes. If I base my ideas on the photographs obtained during the last 6 years, I think that it is better to use the full resolving power of an instrument of only 200–250 mm, less troubled by atmospheric turbulence, than a larger instrument (whose resolving power is reduced, 99% of the time, to about 1″, and sometimes even further). It is also with a 'small' 20 cm that G. Thérin successfully obtains some of the best contemporary lunar photographs. A diameter of 300 mm seems to be the limit, useful to the full only on exceptional occasions (which C. Arsidi and Bernd Flach-Wilken have already shown us). I do not think that it would be worth while to exceed a diameter of 400 mm for lunar photography, even from a very select location. With perfect images, it becomes possible for an excellent telescope to reach details as small as 0″.15, i.e. the equivalent of the best we can obtain from the surface of the Earth.

For planetary photography a 20 cm telescope is plainly insufficient. In spite of the limitations imposed by bad seeing, the best results in amateur planetary photography have been obtained with telescopes of 300–400 mm aperture. Don Parker obtained excellent results with a good 310 mm Newtonian – results further improved after refiguring the main mirror. He exchanged his 310 mm for a 400 mm and his photos were again surpassed in their resolution (because he works in a 'privileged' site). Since he has employed a CCD camera, his results have equalled the best planetary photographs obtained from the Earth's surface. Isao Miyazaki also obtained, with his 400 mm Newtonian, planetary photographs worthy of the best specialised observatories. The choice between 300 mm and 400 mm depends solely on the site. I do not think that having a 500 mm telescope (or greater) would offer any real advantage in amateur planetary photography (even at Pic du Midi, the 1060 mm telescope does not often exceed the resolving power of a 400 mm). In general, the advantage of a giant telescope is its great stability, the quality of its drive and its freedom from wind vibrations. I therefore think, in conclusion, that it is infinitely preferable to use an excellent 300 mm provided with a perfect mounting than a 600 mm of doubtful optical and mechanical quality. Dollfus (1961) has given a good summary of the problem of high resolution photography, at a professional level (in agreement with Gaviola, 1948): the majority of the best planetary images known have an average resolution of 0″.60–0″.40 (which is within reach of the amateur's 300 mm telescope).

2.5 The different types of telescope

Of the different types of telescopes, some are able to perform better than others in high resolution photography. Several are illustrated in Figs. 2.2 and 2.3. We must now review their advantages and disadvantages.

Those telescopes which are entirely of the reflecting type are the most used. The best-known telescope of all is the *Newtonian reflector*. Most importantly, it is also the ideal telescope for high resolution, as its optics are relatively easy to make to a high precision. Many English and American amateurs seem to prefer focal ratios between $F/D = 8$ and 12, in order to reduce to a minimum the diameter of the plane mirror and cause the least central obstruction, facilitate collimation,

Figure 2.2. Some telescopes used for high resolution astrophotography. *1. The 200 mm refractor of G. Nemec (Munich), a modification of Schaer's design, the light-path being reflected by two mirrors. 2. The writer's Starfire refractor (178 mm OG) on a 160P Takahashi mounting (at St. Clément). 3. The writer's Celestron 14 (355 mm Schmidt–Cassegrain). Note the nearly horizontal position of the fork (at Cotonou, latitude +6°). 4. The imposing 400 mm Newtonian of I. Miyazaki (Okinawa).*

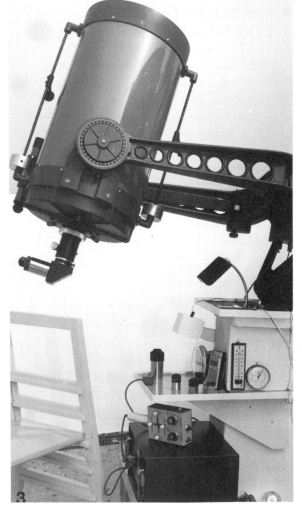

increase the size of the useful field, etc. I think, however, that for a diameter of 300 mm, for example, a focal ratio of 9 or 10 makes the telescope too cumbersome; observing conditions are more difficult and the long tube will be more affected by wind. I am persuaded that the central obstruction of the small plane mirror can be considerably reduced, by introducing an excellent achromatic doublet of the Barlow lens type, just at the entrance to the drawtube. Parker uses one of these transfer lenses, and the constructor Takahashi delivers Newtonians of $F/D = 6$ equipped with two types of transfer lenses: a teleconverter which constitutes a 'planetary' telescope working at $F/D = 8$, and a tele-compressor which transforms the same telescope into an astrograph of $F/D = 4.8$, with correction for coma. Under these conditions, a Newtonian reflector, even working at $F/D = 6$, will only need a small flat mirror, the central obstruction of which will not exceed 0.15 (as in high resolution photography we have no need of a large field). The flat mirror is held in position by a 'spider' made from one, three or four vanes, whose presence entails a modification of the diffraction pattern (introducing diffraction 'spikes'). One can get round this difficulty by closing off the telescope tube with a glass window, which supports the plane mirror at its centre, but the window must be perfectly plane-parallel and fashioned from optical quality glass. Its price greatly exceeds that of the main mirror. A closed tube (with optical window) perhaps has the advantage, in places where the night-time temperature is constant, but it will be particularly detrimental at a site where the temperature falls throughout the night.

The Newtonian telescope gives the viewer the advantage of being able to observe in comfort, especially when the object is close to the meridian. On the other hand, the observation can become difficult when studying an object that is rising or setting (extreme E or W). I therefore think that a Newtonian telescope should have a rotating tube (this is the case with many commercial telescopes: Takahashi, Zeiss, Asko, etc.). The Newtonian offers one final advantage: observation is performed near the upper end of the tube, and it will suffice to use a low, and therefore very compact, equatorial mount. The telescope is also very easy to house and less affected by wind. Concerning turbulence within the tube, I think that a plain tube, perfectly sealed, is preferable to an open one (of the Serrurier type), because of the heat emitted by the observer himself. If the tube is metallic, it will cool down very quickly, while a wooden or stout plastic tube will maintain a constant temperature better, once thermal equilibrium has been attained. Taking too long to easily reach thermal equilibrium, a plain tube may usefully be provided with one or more lateral windows (close to the main mirror) which should be opened an hour or two before observation begins. Alternatively, it can be provided with one or two small electric fans which can help in stabilising the temperature. The improvement given by a glass window (Texereau, 1984) is more illusory than real. The glass has the disadvantage of being very susceptible to dew. It is not at all proven that a glass window is indispensable in obtaining high resolution. The Newtonian telescopes at Pic du Midi, of Miyazaki, Viscardy, Parker, etc., are open to the atmosphere and have amply proved their worth.

Of course, it is of the greatest importance to collimate the telescope. The collimation must be checked before every observation and today there is no excuse for not knowing how to centre the two mirrors, as one finds various commercial devices which ease this task (for example, the 'Lazer-mate' of Astronomical Innovations).

The *classical Cassegrain* telescope can be a good alternative to the Newtonian. On this subject, some strange remarks are sometimes made, such as 'the perfection of the main mirror must be greater than that of a Newtonian, because of the amplification due to the secondary', which is evidently quite absurd. If we are able to accept a cumbersome Cassegrain, we can choose a focal ratio of $F/D = 6$ for the main mirror; its fabrication will be easier (which leads us to hope for high quality), and the central obstruction due to the small secondary mirror can be very small, especially if we wish to have a very long resultant focal length. However, if we wish to have a short tube, we must come down to a focal ratio $F/D = 5$ or even 4. The parabolisation of the main mirror

Figure 2.3. More telescopes. *1. Solar refractor, 110 mm OG, mounted as a refracto-reflector (France; author). 2. 300 mm Schiefspiegler telescope of Bernd Flach-Wilken (Wirges); the main mirror is marked 'm'. 3. 250 mm Newtonian on a fork mount (optics by Bacchi, mounting by Florsch) (author). 4. Heavy Asko mounting supporting Mewlon 250 mm Dall–Kirkham reflector by Takahashi, 130 EDT Apo refractor by Astro-Physics and 60 mm Vixen refractor (author; St. Clément, France).*

will be much more difficult to execute, therefore more costly, and the mirror risks being of poorer performance than one working at $F/D = 6$. The diameter of the secondary mirror must also be larger (and the central obstruction can reach or exceed 0.30). The fabrication of the secondary hyperbolic mirror is hardly an easy task, and this latter component will rarely be close to perfection. Thanks to the amplification of the secondary, we can imagine planetary telescopes with a very long resultant focal length ($F/D = 20$–40), obtained so simply with the two mirrors (from which stems the possibility of working without an eyepiece and therefore gaining in image brightness). G. Viscardy and P. Sette worked under these conditions, but the useful field becomes much reduced and the telescope needs a powerful finder, perfectly collimated. We can even imagine Cassegrains having focal ratios of 80, specially designed for direct photography at high amplification. In spite of their small central obstruction (up to 0.15), these telescopes would be difficult to use in practice. The field of view would be ridiculously small, limited by the very small diameter of the hole in the centre of the main mirror. I therefore think that we must be content with a focal ratio of between 12 and 15. Certain opticians (Texereau, 1984) advocated using an optical window to support the convex mirror, in order to avoid the alteration of the diffraction pattern by a spider, the tube being completely closed. According to some opinions, the Cassegrain telescope, only slightly obstructed and provided with an optical window, should be the ideal instrument for high resolution photography. In fact, this has never been proven. To reduce the diameter of the secondary mirror, one could think of introducing a transfer lens (achromatic doublet) in the optical system, about 12 cm in front of the principal mirror, in order to distance the secondary mirror and thus enable its diameter to be reduced. This is actually now done in the new 300 mm Mewlon Cassegrains by Takahashi, employing the Dillworth catadioptric sytem. In practice, most Cassegrain telescope constructors choose the least favourable formula: a primary mirror with $F/D = 4$ and a central obstruction of 0.30.

There are three types of Cassegrain telescope, as described below.

It is often suggested, particularly in Great Britain (Sidgwick, 1980), that it can be advantageous to use the *Dall–Kirkham* formula, the optics of which are easier to make (elliptical primary and spherical secondary). The inconvenience of this type of Cassegrain lies in its very small field, due to strong coma. The Dall–Kirkham was conceived entirely for lunar and planetary observation (or for double stars), at high power.* With one of these telescopes, working at $F/D = 10$, it is not possible to photograph the whole of the Moon (the actual field does not exceed 25′). In spite of this inconvenience, Takahashi, one of the most famous Japanese constructors, did not hesitate to construct a whole series of Dall–Kirkham Cassegrains. Under the name 'Mewlon', these telescopes are built in three sizes: 180 mm, 250 mm and 300 mm. The Mewlon 250 has the following characteristics: total diameter of primary, 260 mm; useful diameter, 250 mm; $F/D = 12$, leading to a focal length of 3 m; secondary mirror diameter, 72 mm (central obstruction, 0.34). The primary seems to be about $F/D = 4$. The Mewlon 250 is a very carefully made telescope, the focusing is by electrical movement of the secondary, and the interior of the rear baffle tube contains about a dozen diaphragms. The first trials, on the sky, showed an excellent optical quality and very good contrast. We can ask, therefore, why Takahashi is starting production in such a big way with telescopes of this type (specialised instruments for the observation of objects of small angular diameter). I think that it is because the production of the two mirrors is much easier than in all other complex telescopes. The manufacturer is therefore hoping for excellent optical quality at a competitive price.

It is the classical Cassegrain which is the most used by amateurs. The fabrication of the mirrors is more difficult (parabolic primary and hyperbolic secondary). It is therefore more expensive to make, and the end result is not always perfect. The classical Cassegrain benefits from a larger field, close to the same order of size as a Newtonian of the same focal ratio. Aberration due to coma is lower in the classical Cassegrain, but astigmatism and field curvature remain considerable.

The *Ritchey–Chrétien* design is most often used in the large observatories. Difficult to produce (both mirrors are hyperboloids!), this type of telescope offers the advantage of a very low coma,

---

* Using a 400 mm example of the telescope he himself had designed, the late H. E. Dall of Luton, England, built up a reputation for his fine planetary photographs, especially of Mars and Jupiter. He obtained some images of Jupiter in the mid-1970s which for their time were of remarkable quality, using Tri-X and similar emulsions. (RJM)

allowing the use of wide fields (up to 0°.7), but astigmatism and field curvature require the use of correcting systems if one wishes to have a useful field wider than 1°. Given the difficulty of making the mirrors, its high price and the uncertainty about the perfection of the optical system, it is best to reject the Ritchey–Chrétien for high resolution photography.

In practice, the Cassegrain is rarely as good as the Newtonian, as a result of the low focal ratio of the primary (which is therefore hard to parabolise precisely), coupled with the difficulty of making the secondary. Its only advantage lies in having its focal plane just behind the primary mirror, making for very comfortable observation, even when observing far from the meridian. A disadvantage of these Cassegrain telescopes is the fact that they look directly at the sky, through the hole in the main mirror, which leads to a phenomenon known as 'sky-flooding'. From this there follows a loss of contrast (if the sky is so much as even slightly luminous), unless we enclose the convergent beam from the secondary mirror in a very long tube, mounted upon the opening in the main mirror and concentric with the optical axis. The length and diameter of the protective tube (sometimes called a light baffle) must be worked out exactly and the tube provided with diaphragms. All light which does not contribute to the formation of the image must be eliminated by this device.*

The standard *Schmidt–Cassegrain* telescope (Celestron, Meade) was invented by R. R. Willey, Jr., in 1962. The primary mirror works at a very fast focal ratio ($F/D = 2-3$) and has a spherical form. The secondary is convex and slightly ellipsoidal. An aspheric Schmidt window corrects the enormous spherical aberration of the ensemble. Normally, the Schmidt correcting plate must be placed at the centre of curvature of (at twice the focal length from) the main mirror. To shorten the length of the tube considerably, the Schmidt plate is placed very close to the focus of the main mirror, which allows the secondary mirror to be fixed by means of a hole pierced in the centre of the corrector plate itself (thus eliminating the spider). The unconventional position of the corrector plate leads to a strong curvature of the field, clearly explained by

Rutten & van Venrooij (1988). Theoretically, the Schmidt–Cassegrains at $F/D = 10$ present a smaller coma than a Cassegrain with the same characteristics. It is unfortunately the curvature of the field which prevents photography of the whole Moon at high resolution (useful sharp field, 30′). This field curvature can be eliminated by moving the Schmidt plate to its normal (2F) position, as the astrophotographer Kim Zussman did, or by refiguring the secondary mirror (which is what is usually done, with a poor degree of success). Initially, the Schmidt–Cassegrain was no more than an optical curiosity, since its production presented problems. It was following a Patent of Johnson, allowing the Schmidt plate to be polished in a completely automatic manner, that Celestron were able to launch into industrial production of Schmidt–Cassegrains, followed – very rapidly – by Meade, Criterion, Bausch and Lomb, and then Takahashi. Although the specialists were little inclined to admire the new telescopes, the Schmidt–Cassegrain enjoyed enormous success in the hands of a great many amateurs, tempted by their compactness, their lightness, and their ease and comfort of use. Today these telescopes are among the best sellers and there must be more than 200 000 of them across the world! What value are they, really? Opinions are very mixed. One can immediately criticise the complexity of the optics and their lightweight mountings. Martinez (1987) did not hesitate to write, 'consequently Schmidt–Cassegrain scopes have primary ratios that are too small (F/D about 2.5) and obstructions that are too large (0.30–0.38) to provide images with enough contrast for high resolution. In fact, the resolving power of these scopes, for objects low in contrast, is that of a perfect instrument of half the diameter.' That is quite an extreme opinion, since we know today that some owners of Schmidt–Cassegrains of only 200 mm aperture have been able to take the most extraordinary high resolution photos (see Figs. 4.17, 4.19, 4.20, etc.). The author (Dragesco, 1978) published tests of Celestrons 5 and 8. The diffraction patterns were relatively correct (as good at focus as extrafocally), but the Foucaultgrams showed numerous small-amplitude defects. In addition, each telescope tested was slightly different. The mountings were too lightweight, the drive accuracy was low and the sensitivity to vibration was intolerable (the shutter release of a reflex camera caused a vibration of 3″.8, which could not be overcome for exposure times longer than 1/250 s). It is therefore scarcely credible to read, in the manufacturer's

---

* The 1060 mm Cassegrain at Pic du Midi must have either no light baffle or a very short one, for in my experience (in 1986) it was hard to follow a planet far into daylight before it was lost against the bright sky background. In contrast, with the Great Meudon refractor one could follow a planet like Jupiter for hours into daylight if one wished. (RJM)

instructions, that it will suffice to attach a 24×36 reflex camera to a Celestron to be ready to take good high resolution photos. Like all small telescopes, the Schmidt–Cassegrains can only be used for solar photography (if possible, at $1/1000$ s) with reflex cameras, or, for exposures longer than $1/125$ s with the aid of a Compur-type shutter or by making use of the 'hat-trick' technique (see p. 39). Di Cicco (1989) carried out a precision study of the optical components of several 200 mm Schmidt–Cassegrains made by Celestron and Meade. Photographs using the autocollimation test showed faults of the various surfaces, which did not permit him to conclude that he was dealing with optics of high quality. That is quite normal and logical: it is impossible to produce, at a competitive price, a perfect spherical mirror, working at $F/D = 2.5$; a perfect Schmidt plate (fashioned from normal window glass rather than borosilicate optical glass by an industrial process*); and a small convex mirror of high quality, manually aspherised. A complete system could be perfectly made by an optician, refiguring each piece by hand, but that would lead to a prohibitively high price. So, without being really bad, Schmidt–Cassegrain optical systems remain of average quality, variable from one telescope to another. One can find, by chance, a high quality Celestron 8, but also a C14 with defects as large as 2λ (!). It was thanks to a Celestron 8 that G. Thérin was able to obtain lunar photos showing the highest resolution ever attained on a celestial body by a telescope of that diameter (see Fig. 4.21). These photographs are even better than some of the finest ones taken with the aid of the most famous telescopes in existence. It is all the more disconcerting as, according to theory, the central obstruction of 0.34 of the C8 should limit its ability to reveal fine planetary details, which is evidently not the case. Therefore, we must refrain from placing more than limited confidence in the underlying theoretical reasoning!

For 3 years I tried to obtain lunar photographs with a C8 at Cotonou with near-perfect images. I was unable to obtain good ones, owing to the imperfections in the drive. At that time (1977–81), only Arsidi had succeeded in obtaining good lunar photographs, by repeatedly adjusting the drive of his C8. Later, still at Cotonou, I was engaged in a programme of high resolution photography, with the aid of a Celestron 14 (355 mm diameter). The optics were not very good, but thanks to the excellent local seeing I was able to obtain some hundreds of lunar and planetary photographs which were considered to be very good for the time (1980–84). They were widely published, and several of them were awarded prizes (*Astronomy* magazine, BAA, Orange County Astronomers). Today they have been completely surpassed, their average resolution being about $1''$ (the resolving power of a 13 cm).* Only two of the lunar images reached an acceptable resolution (by today's standards): $0''.40$ on the rilles of Stadius and $0''.27$–$0''.30$ on the rille of the Alpine Valley. Therefore, the limitation was not due to the optics (although $0''.30$ could be reached with a good 20 cm!) but to the drive. I have written (Dragesco, 1984) of the trouble I gave myself in continually adjusting the drive of my telescope, with the help of an excellent variable-frequency Astrotech system (quartz-controlled). It was necessary to adjust the speed of the drive (with the help of an eyepiece graticule) every 3–4 photos. The image was moving continually in RA, one way or the other (in an erratic manner), such that any exposure longer than 1 s was penalised by a loss of resolution. But at Cotonou, owing to the persistent haze, the exposure times were from 2 s to 4 s. Since then, the majority of astrophotographers owning Schmidt–Cassegrains have ended up by remounting the tube upon a more rigid mount, provided with a better-quality drive. Thomas Mathis & Co. offer heavy fork mounts, the 500 model being suited to the C14; Byers also produce a mount and a drive for the C14. G. Thérin uses his C8 on a heavy mounting, the Vixen 106. When CCDs come into general use, which require exposure times of only $1/10$ s on the Moon or Jupiter, problems with the quality of the drive will matter very little. In the same way, for the last year or so, Celestron as well as Meade have entirely revised their telescope drives and improved the rigidity of the mountings. Nevertheless, the Schmidt–Cassegrain remains an interesting instrument, despite its various inherent defects!

R. D. Sigler (1974–5) conceived a complete series of improved Schmidt–Cassegrain telescopes, which were of great interest, but their high price

---

* Now the corrector plates are made from borosilicate optical glass.

* In spite of later advances in photographic technology, these planetary images from the early 1980s (and the earlier amateur ones by Němec, Viscardy and Dall), remain of great value as *contemporary documents*, which are sometimes the best available for a particular planet for a particular date. Even today, in my opinion, they would count as fine examples of the best lunar and planetary photography. (RJM)

of manufacture was perhaps prohibitive. The best solution could be that of Takahashi, who are introducing a 225 mm Schmidt–Cassegrain, with a higher F/D for the main mirror, provided with a good mount. The price is necessarily a little high, but remains a possibility. The first trials of this instrument have shown great promise. The Japanese firm Asko are introducing Schmidt–Cassegrains of very high quality in the following diameters (mm): 210, 230, 260, 300, 350 and 400. The prices of the tube plus optics alone are very high, twice as much as that of their conventional Cassegrain of the same size! (A 210 mm costs ten times as much as a C8.) The equatorial mountings are of a professional standard.

For some Central European observers, the best formula for high resolution is to acquire a *Schiefspiegler* (as designed by A. Kutter), or more specifically a tri-Schiefspiegler, with three mirrors. These telescopes, thanks to inclined mirrors, benefit from the absence of any central obstruction and long focal lengths (Fig. 2.4). They have given excellent results in astrophotography in the hands of B. Flach-Wilken and T. Platt. The production of their specially deformed mirrors is a very delicate task, so they cannot therefore be mass-produced. The firm of Lichtenknecker can make the optics, to the specification of the client. Considered as the equivalent of a refractor, the Schiefspiegler is often badly handicapped, owing to imperfections in its optics (the manufacture of

which, as already noted, is difficult) and the difficulty in its collimation. I doubt that this type of telescope is destined to have a great future, in spite of the enthusiasm of owners.

The *Maksutov* telescope (1944) is also a catadioptric of the Cassegrain variety: it is to Gregory (1957–8), that we owe the Maksutov–Cassegrains of $F/D = 23$ and $F/D = 15$, which have enjoyed a certain popularity. The Gregory–Maksutov is a short and compact telescope, and its spherical surfaces ($F/D = 23$) perhaps facilitate its fabrication: spherical perforated primary mirror, spherical meniscus corrector and an aluminised spot at the centre of the latter serving as the secondary mirror. Owing to the thickness and the strong curvature of the corrector, the Maksutov has a low chromatic aberration (negligible for focal ratios between 10 and 23).

The Maksutov–Cassegrain telescope has enjoyed success, particularly by virtue of the instruments manufactured by the firm of Questar (USA). In spite of its small size, the Questar Standard (3.5 in; 88 mm) is very well made, optically as well as mechanically, and the finish is impeccable. Although very expensive, the Questar 3.5 in has sold by the thousand, and remains well used, especially in the USA, in the teaching of astronomy in colleges, and also as a portable instrument (which is easy to take on an eclipse expedition, etc.). However, it is no more than a small telescope of 9 cm aperture, which does not permit of any really serious work from our point

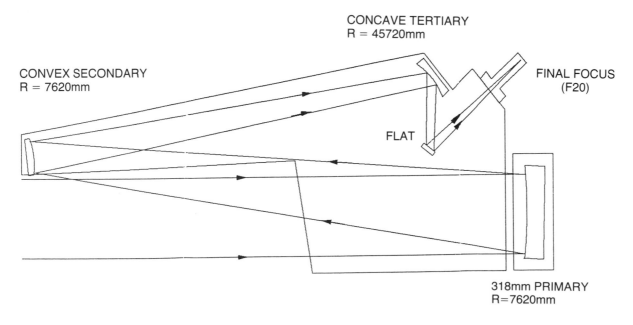

GENERAL VIEW OF QUAD-SCHIEFSPIEGLER OPTICAL SYSTEM

Figure 2.4. General optical diagram of the quad-Schiefspiegler *of Terry Platt, 316 mm aperture. This telescope, having no central obstruction, gives very good planetary images.*

of view. The Questar 7 (178 mm) is a much more interesting instrument, capable of giving excellent results in solar photography, but again too small for high resolution lunar and planetary work. Its price is really high. The Questar 12 (300 mm) is an exceptional instrument (although not particularly designed for high resolution) but its price is truly astronomical. Carl Zeiss Jena are introducing a new Maksutov–Cassegrain, the 'Meniscas 180', with $F/D = 10$. The central obstruction is lower than in the equivalent Schmidt–Cassegrain and the finish is extremely neat, but the equatorial mount seems to me to be much too lightweight. In spite of its relatively high price, the Meniscas 180 cannot really be cosidered as a high resolution telescope, notably because its aperture is too small.

The *refracting telescope* has always been considered as being the ideal instrument for high resolution observation (double stars, planets, Sun and Moon). Refractors are simple. Their objectives have spherical surfaces (which are not too difficult to make); there are no awkward effects due to central obstructions or spiders and no scattered light from aluminised surfaces; the optics tube is closed, and the collimation is easy and does not need further attention.*

However, refractors remain very little used in high resolution work, because of two great drawbacks: they are expensive and very awkward to use (and also often possess residual chromatic aberration). In spite of these inconveniences, there is a real 'refractor culture', notably in Great Britain and Japan. The first superb lunar and planetary amateur photographs were obtained in 1964–5 by Günther Nemec (Fig. 4.16), whom I was able to meet in Munich. His coudé refractor used a double achromat by Lichtenknecker of 200 mm diameter, $F/D = 20$. The photographs

obtained by Nemec were really surprising for their time (but easily surpassed today). Their average resolution rarely reached 0".40, which is low by today's standards. The great inconveniences of refractors remain in their residual chromatic aberration (or secondary spectrum) and the great length of their tubes (3 m for a 20 cm $F/D = 15$ objective). Great progress has been made during the last 10 years, and the modern refractor is slowly commanding more and more respect. The use of computer technology, new optical glasses of low dispersion and synthetic fluorite facilitates the fabrication of shorter refractors and nearly perfect apochromats (Fig. 2.5). Although practically all the short modern refractors are called apochromats, the term is often employed in a misleading way. From the number of their lenses (2–4), the nature of the glass employed, their diameter and focal ratio, these new objectives fall into one of the following categories: improved achromats, semiapochromats or even high performance apochromats. Of course, the prices of these telescopes are relatively high, but often lower than the sort of $F/D = 15$ three-lens apochromats which have been made for the last 25 years. I have had a certain amount of experience with a Vixen 102 mm $F/D = 9$ apochromat. This objective yields practically perfect images but its resolving power remains that of any 102 mm lens, good enough for visual observations but, photographically, inferior to that of a good 150 mm Newtonian reflector (very much cheaper). The refractors of fluorite or ED glass (doublets for small apertures, triplets for larger ones) made by Vixen, Meade, Takahashi, Astro-Physics or Carl Zeiss Jena are usually excellent and therefore very suitable for high resolution solar photography (diameters 60–150 mm). Amateurs such as C. Arsidi and C. Ickhanian have shown that it is possible to take excellent lunar photographs with the Vixen 102, and with a resolution close to the theoretical limit, but the telescope is too small to be able to reach truly high resolution. There is much to commend about three-element objectives, more or less apochromatic. We already know that the Takahashi 125 and 150 are excellent (but very costly) and that the Vixen 130 and 150 are very good also, and the same applies to the new Astro-Physics EDT (Fig. 2.5). The new Carl Zeiss Jena 130 and 150 are instruments of extremely high performance. Working at $F/D = 8-9$, these triplet lenses give quasi-perfect images but remain small in diameter. Astro-Physics and Meade offer us 180 mm refractors with doublet or triplet

---

* Zmek (1993) has published a meticulous comparative study of the faults of reflectors versus refractors. If all the factors studied by the author are taken into account, one can arrive at the conclusion that, in a good many instances, a 100 mm refractor is able to show the same planetary details visually as a 200 mm reflector, which is evidently untrue. By only considering the central obstruction factor, Zmek asserts that a perfect refractor (of which there are but few) of a given diameter allows the same planetary details to be viewed as with an excellent reflector whose diameter has been reduced by an amount equal to the diameter of the secondary mirror. Thus, my 250 mm Cassegrain, which has a 72 mm diameter secondary, should give the same planetary resolution as a 178 mm refractor. In fact, the Cassegrain should fare even worse, for it will be handicapped to a greater extent by atmospheric turbulence. It is therefore interesting to note that I possess a 178 mm apochromatic refractor, and that I have compared its performance with that of my 250 mm Cassegrain. I consider that the Cassegrain is both more convenient to use and has a higher *photographic* resolution than the 178 mm refractor. Once again, theory and practice disagree, above all in comparing visual observation and photographic use.

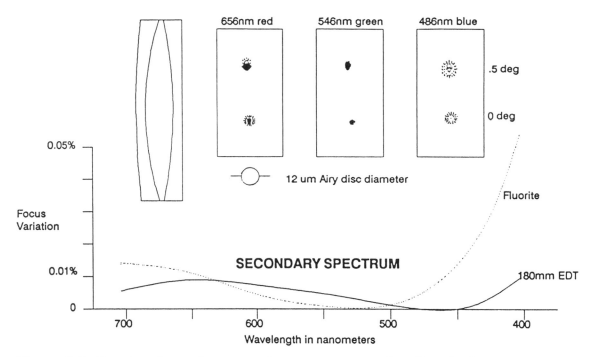

Figure 2.5. Showing the performance of an EDT apochromatic triplet objective *by Astro-Physics, giving its secondary spectrum (focus variation against* *wavelength) and schematic drawings of its diffraction images, obtained at three different wavelengths, for a field of view of 0°.5.*

object glasses; Takahashi, Goto and Zeiss 200 mm apochromats ($F/D = 8$ or 10). These instruments can aspire to high resolution lunar and planetary work. Theoretically, for an equal diameter, an apochromatic refractor should be superior to any type of reflector: its optical simplicity guarantees very crisp images with high contrast, as the closed tube and the absence of a central obstruction seem to reduce the unpleasant effects of astronomical turbulence. However, a diameter of 200 mm is a little low for planetary photography. The price of these refractors (notably those with fluorite triplets) reaches the highest levels and their length (2 m for a 200 mm $F/D = 10$) remains prohibitive. I therefore think that, for good images, we always have the option of using a high-quality 250 mm or 300 mm Newtonian. It would be unrealistic to think of an amateur using a 300 mm refractor: it must be apochromatic, its length would exceed 3 m and its price would be four or five times higher than that of the best reflector of the same diameter, without any real advantage over the latter.

For those who possess only an achromatic refractor, we can try to eliminate the secondary spectrum thanks to a combination of film and filter allowing us to work in nearly monochromatic light. For solar photography (at Pic du Midi) it was Rösch (1959) who used an orthochromatic film and an orange filter ($\lambda = 595$ nm). We can very easily reduce the secondary spectrum with the help of a Wratten 53 green filter, but TP 2415 film has a low sensitivity to green light: useful, then, for solar work.

Finally, if the refractor remains the ideal instrument for observing and photographing the Sun (in white light, in hydrogen alpha or in use as a coronograph), the Newtonian or Cassegrain from 250 mm to 400 mm aperture forms the basic instrument for lunar and planetary photography; we are waiting for a $F/D = 8$, 250–300 mm refractor at a reasonable price!

I had the chance to observe and photograph with the great old-fashioned refractors (of 60 cm and 83 cm diameter) and I was very disappointed. These instruments have an enormous secondary spectrum and their mountings are too unstable. To conclude the optical side of things, I must recall that, although I have already owned 32 telescopes, I never obtained images so perfect as those I achieved with a 260 mm Newtonian (the mirror of which, polished by Bacchi, was accurate to better than $\lambda/30$). That said, we can obtain excellent high resolution photographs with any sort of telescope, provided that the optics are good, the site is a favourable one and the photographer is competent!

2.6 Telescope mountings

The mistake of the beginner is to be hypnotised by the objective (mirror or lens), to spend the most money on the optical tube (of the largest diameter possible) and to neglect the mounting. This is a grave error in photography. Better to have a good small objective on a robust mounting than a large objective on a badly designed flimsy mounting. The majority of small refractors (60–90 mm) never allow us to obtain photographs as good as their theoretical resolving power, because of their inadequate mountings. It is, even today, the worst error of the manufacturers, who seek to sell their instruments with as large a diameter as possible, at a competitive price: they always economise on the mountings. For visual observation, this is not always an inconvenience, but for photography (at high resolution or for long exposures) it is catastrophic. One must accept having to spend as much on the mounting as on the optical tube. Besides, practically all the great celestial photographers, or those practising high resolution, most frequently use commercial telescopes (including the Schmidt–Cassegrain), but install them on larger, heavier mountings provided with excellent drives. When building the telescope him-/herself the amateur must never forget to make the tube large enough (some 10 cm wider than the main mirror) and completely closed (no Serrurier-type tubes for high resolution!). The tube of a Newtonian telescope should be capable of being rotated freely, or else provided with several places for the eyepiece: meridian, east or west. The mounts of the earlier Schmidt–Cassegrains were very portable but too lightweight. Celestron and Meade seem to have understood this and have sensibly strengthened their current mountings, although more work needs to be done in this field. Takahashi appear to have a very good solution to this problem with their SCT 225. In a general way, it is always preferable to buy a heavier mounting than is recommended by the manufacturer. Mounting, for example, a Celestron 8 on a Vixen Super-Polaris mount is an economical solution but an insufficient one. The mounting is still too light. The Vixen 106 is much better. Even if portable, a mounting must be heavy and solid enough to resist a gentle breeze; the optical tube must not vibrate at the moment when the hand of the observer is adjusting the focus. When it is necessary to have a portable mounting (as it is often useful to look for more favourable observing sties), it is essential to have a precise polar alignment, even if we have to fiddle about with the adjustment by Bigourdan's (or similar) method. One of the major faults of most Schmidt–Cassegrains is the absence of a polar sighting tube. The goal of high resolution photography necessitates precision alignment; an error of two or three degrees is unacceptable, even with exposures of two or three seconds, if one demands the theoretical resolving power of a 20 cm telescope. A polar axis misaligned only by 1° leads to an error of 1″ every 4 s of time (Martinez, 1987).

The equipment used for driving the telescope plays a critical role. The majority of commercial telescopes, as well as those constructed by most amateurs, use the classical system: a toothed wheel driven by a tangent, or 'worm' screw. The precision of a given drive system increases with the diameter of the wheel. Dobbins et al. (1988) consider that this wheel should have a diameter at least as great as that of the main mirror, as the great professional reflecting telescopes often have driving wheels twice as wide as the mirror. All the commercial instruments have toothed wheels which are too small. We think, in practice, that the precision of the drive depends, above all, on the quality of manufacture of the mechanical components and therefore a 75 mm wheel could be made to a higher quality than one twice or three times as large! We must not be misled: better to have a small precision-made drive than a large but badly made product.

We can and must test the precision of the drive of the instrument that we wish to use for high resolution photography. The method most frequently used (also described by Wallis & Provin, 1988) consists in displacing the polar axis by 5° in azimuth, then photographing a star situated on the ecliptic, for about 30 min. We thus obtain a sinusoidal trace (rather than a straight line). By applying the correct theory, the deviations can be measured to give a good idea of the quality of the drive. The significant deviations are due to a radial fault in the worm, and the small deviations to irregularities in the cutting of the teeth in the large wheel itself: they can perhaps be confused with those caused by bad seeing. On the latest Schmidt–Cassegrain models electronic gadgets cleverly alter the speed of the drive motors, as a function of the periodic error due to mechanical errors (for example, if the worm is not properly centred on its rotation axis). For high resolution photography, the most important test is the one that I have always used and which is also given by Wallis & Provin (1988). For its execution it is essential to have a high power eyepiece, provided with a fine graticule, such as the Takahashi eyepiece of 3 mm focal length, or a 12 mm Plössl, used with a Barlow lens. For a 200 mm telescope, it is necessary to observe with an eyepiece with a magnification of about ×300 (×400, if possible). It is best to choose the Moon rather than a star to observe. We search for a minute, well-contrasted craterlet (two or three kilometres in diameter), at the centre of the graticule (for which it will be absolutely necessary to use good slow motions in RA and declination). The motor being run at the lunar rate, the craterlet should remain at the centre of the graticule. In reality, it will move. The problem is rendered more complex by virtue of the atmospheric turbulence, which has a tendency to displace the craterlet in all directions. We must concern ourselves solely with what happens in alpha (RA): the turbulence may make the craterlet dance, but it must constantly return to the cross on the graticule. Thus, a slow drift in RA will quickly be seen. It will be caused either by poor adjustment of the variable-frequency drive, by the control box governing the 'lunar rate', or by mechanical or electronic imprecision in the drive mechanism. To have an idea of the seriousness of the fault, we must calculate the drift, in arcseconds per second of time. A craterlet measures about 1″ or 2″. For it to be visibly registered on a sensitive emulsion it must stay perfectly still throughout the usual exposure time (of 1–3 s). If, during 2 s, the craterlet is displaced by twice its diameter, it is clear that it will not be recorded on the negative (or only as a hazy, elongated spot). We must now look to improve the precision of the drive, by tinkering with the variable-frequency control or by modifying the electronics which control the speed of the 'lunar' rate, on the control box of the stepping motor. However, it is also possible to have imprecision due to imperfection in the mechanism. In this case we are unable to attempt to establish the maximum permissible exposure, on account of the likely drift of our instrument. Often we are obliged not to exceed an exposure time of 1 s, which is very limiting, and leads us to use low focal ratios (causing a loss of resolution, because of the granulation of the film). We must repeat the same experiment with a planet – Jupiter, for instance. With the instrument driven at the 'sidereal' speed, the planet should stay at the centre of the graticule, in spite of any irregular movements due to bad seeing. There also, if the planet drifts (to the E or W) more than 2″ (that is, more than 1/20 of Jupiter's apparent diameter) during 2 s of

time, then the photographic resolution attainable will hardly exceed 2″.

Many people underestimate the effect that a faulty drive can have upon a high resolution photograph. If our colleagues Miyazaki and Parker have been able to arrive at very high resolution photographs, with exposure times of 3 s or more, that proves, more than the excellence of the site, the quality of the drives of their telescopes. In case of doubt, the best thing is to shorten the maximum exposure time to 1 s or to 1.5 s. One of the advantages of CCDs lies, justly, in their very high sensitivity, with the possibility of using exposure times of 0.2 s or even less: such exposure times render useless the study of the precision of the drive.*

For some years several Japanese manufacturers have produced very beautiful equatorial mountings, equipped with excellent stepping motors but with control boxes that lack more than two drive speeds: sidereal (stellar) and solar (quite useless). The lunar rate is more important; perhaps photography of the Moon is of no interest to the engineers who designed the control boxes! In contrast, American telescopes (notably Celestron and Meade) have access to all possible drive rates. However, the Moon also has a motion in declination, reaching 0″.26 per second of time. Therefore, to be able to exceed an exposure time of 2 s, most desirable in obtaining high resolution photographs, one must think about making an automatic correction in delta (declination). The German amateur G. Neṁec had foreseen this and put it into practice (witness the low sensitivity of photographic emulsions in 1965). We can attempt to modify the control boxes by using the correction in declination (towards the north or towards the south, following the movement of the Moon herself), by adjusting the potentiometer to eliminate the drift in declination. (Again, the eyepiece graticule will be needed.) Otherwise, we must be careful not to exceed an exposure time of 2 s.

French amateurs use, on a very wide scale, the driving technique called a *secteur lisse, ruban et écrou* (a 'screw and sector drive' in English), the motion being imparted by a long, endless screw. The device, easy to make and very precise, has

---

* Although the exposure time of a CCD camera can be very brief, there arises the problem of keeping the image within the very tiny field of the CCD camera for long enough to make a series of exposures. Thus, some devotees of this new field consider that the precision of the drive has become more important rather than less so! (RJM)

been described by Texereau (1984), who gave it the above name. Verseau (1986) has compared the precision of movement of this system with that offered by commercial worm and wheel systems. If the radius of the sector is large (up to 50–60 cm) and the screw and its nut are accurately made, it is possible to make a system whose driving accuracy is very much better than that of a commercial instrument. Nevertheless, there are two factors which overshadow the technique: (1) the assembly is bulky, and therefore awkward, and (2) the duration of the driving action is limited (up to 2 h on average, which is not too generous). It is very surprising to find, in spite of the publication of Texereau's book in English, that this type of drive remains almost unknown in the English-language periodicals. This is not the place for me to describe the device, but the explanations given by Texereau (1984, pp. 244–69) are sufficiently detailed and the system's great efficiency has been amply proven.

For high resolution photography, a precision drive equipped with a correction for the lunar rate is not sufficient. It must also have precision movements in alpha and delta, to allow for exact centring in the field of the camera. In the case of stepping motors it is relatively easy, the control boxes providing accelerations up to a factor of ×8 or more, in alpha as well as in delta. In the case of a synchronous motor with a fixed drive rate, one must be able to have a practical system of clamping, to make use of a supplementary 12 V

DC motor (or with AC, as for the C14). In declination, the best system remains that of the screw thread, controlled by a variable-speed 12 V motor. With these two gadgets, there is no time wasted, even when changing direction from north to south. Unfortunately, most Japanese manufacturers have adopted the worm and wheel, a system which always exhibits some inertia when the direction is reversed (due to the space, albeit small, between the engagements of successive teeth). It is not very suitable for high resolution photography.

To conclude, high resolution photography necessitates a rigid mounting, perfectly set up (polar axis parallel to the Earth's to within 1°), equipped with a precision drive, capable of being varied by different, but constant, amounts (with, if possible, quartz-controlled stepping motors), also with rapid motions in RA and declination, necessary for centring the object to be photographed. Enormous progress has been made; the present stepping motors have a tiny consumption of electricity and work off small 12 V batteries (or Ni–Cd accumulators). Personally, I have been as pleased with my two Takahashi 160P mountings as with my Asko mount. With the introduction of CCD receptors, the precision of the drive has lost its former importance but, because of the very small surface area of these devices, slow motions in RA and declination are absolutely indispensable.

Photography at the telescope

3.1 Enlarging the prime focus image

To be able to photograph solar, lunar or planetary details, by making use of the intrinsic resolving power of a given telescope, it is necessary to adjust the focal ratio F/D to enlarge the image as a function of the resolving power of the film. The image that is obtained at the focus of a telescope presents a maximum resolution that is inversely proportional to the focal ratio (Table 3.1).

The resolving power of the film varies between 60 and 300 lines per mm. Gradually, as the limit of resolving power of the telescope is approached, the contrast tends towards zero. Therefore it is necessary to enlarge the prime focus image adequately, as, at the limits of the resolving power of the film, the contrast is considerably lowered. The final focal ratios, those most generally used, vary from 50 to 150 according to Rouse (Wallis & Provin, 1988), from 45 to 100 according to Dollfus (1961) or from 70 to 200 according to Parker & Miyazaki. In fact, all depends on the film employed. With the 'miracle' film Kodak TP 2415, the resolving power of which is very high (but varies greatly with the type of development), focal ratios as low as 50 will suffice, with little loss of resolution. On the other hand, if the film is a rapid infrared one, of low resolving power, it is necessary to raise the focal ratio F/D to 100 or even to 150. On this subject, we must point out that two contradictory tendencies are found: some use very long focal lengths; others use distinctly short ones. The two best-known planetary photographers have opted for the maximum amplification: $F/D = 100–200$! They justify their choice by a desire to enlarge the negative as little as possible, in order to obtain the highest possible photographic quality (considering that at $F/D = 200$ the resolution of the image projected on the film is at most 5 lines/mm and that TP 2415 film easily resolves to 100 lines/mm, even for low contrast, the 'margin of safety' seems to me to be enormous!). The inconveniences of very long focal lengths (80 m for a 400 mm telescope!) are several: (1) the exposure times are very long (from

Table 3.1 *Resolving powers obtained with various focal ratios F/D*

F/D	20	40	70	80	100
Resolving power (lines/mm)	94	47	26	23.5	18

3 s to 8 s or more) and poorly adapted to the constant movements of the image due to bad seeing or to errors in the drive; (2) the field is tiny and necessitates the use of two finders, the more powerful of which needs a high magnification of, say, × 30; (3) the image is very enlarged on the focusing screen of the 24 × 36 reflex camera, leading to real focusing problems. However, I have to confess that those who have chosen this method have obtained magnificent results. This technique suited them, for their sites have low turbulence and their telescopes have excellent drives. The majority of high resolution photographers, and notably the French specialists (G. Viscardy, C. Arsidi, G. Thérin and myself), prefer a smaller enlargement of the primary image, contenting themselves with the range $F/D = 50–100$ (with an average of 60–70 for the Moon, 70–80 for Jupiter, 100 for Mars). These astrophotographers have considered that under their conditions of work an exposure longer than 2.5 s leads to a loss of resolution. The resolving power of TP 2415 film is so high that a very great enlargement does not result in any major problems: no coarse grain, no loss of contrast or resolution. One example will demonstrate the truth of this reasoning: I was able to obtain, in July 1986, an exceptional photograph of the planet Mars (Fig. 4.38) with the 106 cm telescope at Pic du Midi, and one of the best ever taken for its time (as was acknowledged by the leading specialists). However, the image on the negative was less than 5 mm across (F/D about 50). I reached the same conclusion on the subject of an exceptional Jupiter photograph (Dragesco, 1986; Dragesco & McKim, 1987; McKim & Dragesco,

1987): see Fig. 4.44.* It is not at all certain that I would have achieved higher resolution by using a higher focal ratio. In 1973, when I was working at the Pic du Midi with C. Boyer, with the same 106 cm telescope, but with Ilford Pan F film with a much lower resolving power and insufficient contrast, we were using a focal ratio F/D of 92. To conclude, I would say that it is for each individual to discover the technique which is best for him. I advise beginners to use the following focal ratios (with TP 2415 film): Moon, 50; Venus, 50; Jupiter, 70; Mars and Saturn, 80.

Once the focal ratio has been decided, we must find the best technique for enlarging the primary telescopic image. If from the start we have a very long focal length (Cassegrain telescope, $F/D = 24$, for example) we can be content to use a simple Barlow lens, a negative achromat, to multiply the focal length two- or threefold. The small Clavé Barlow is excellent and inexpensive, its negative focal length being 112 mm. A doublet negative Barlow lens is, in principle, corrected for spherical aberration for a well-defined projection distance (22 cm for the Clavé Barlow). In fact, we can increase this distance, without much trouble, up to 2.5 times the amplification of the primary image (so that a $F/D = 24$ Cassegrain becomes a $F/D = 60$ system). One can easily find, by experiment, the exact position of the Barlow lens for a given amplification. One can also calculate the numerical data which are being sought, after formulae given by Sidgwick (1980). Taking $\mathcal{F}$ as the resulting focal length, F as the focal length of the telescope, f_b as the focal length of the Barlow lens and d as the separation between the Barlow lens and the focal plane of the telescope, the resulting focal length $\mathcal{F}$ becomes

$$\frac{F \times f_b}{f_b - d}$$

the amplification factor of the Barlow, or M_b, becomes

$$M_b = \frac{\mathcal{F}}{F} = \frac{f_b}{f_b - d}$$

and the distance, d, from the Barlow to the new focal plane becomes

$$d = f_b (M_b - 1)$$

There are also ×3 achromatic Barlows (that of Clavé is 50 mm in diameter), but these doublets remain difficult to use with amateur instruments, as the distance between the original focal plane and the new one becomes large. At Pic du Midi in 1973 for some time previously (on the advice of J. Texereau), a combination of two achromatic ×3 Barlow lenses coupled together had been used, to obtain a focal ratio $F/D = 92$, with the 106 cm telescope. Martinez (1987) has suggested the use of two ×2 Barlow lenses but he imagined them to be well separated.* In fact some of the best French astrophotographers have already used two coupled ×2 achromatic Barlow lenses. I was tempted to experiment with various projection distances, on average from 16 cm to 20 cm; focal ratios of $F/D = 40$–70 could easily be attained (from the initial focal ratio of $F/D = 12$), with a very bright image and a good sharpness over the whole field (without, however, eliminating the curvature of the field of certain Schmidt–Cassegrains). It is an excellent plan, useful with Cassegrains, Schmidt–Cassegrains or refractors. For Newtonians with $F/D = 5$–7, recourse must be had to more powerful, more compact systems. The camera, being attached to the side of the tube, tends to unbalance the telescope when it is far from the focal plane. In every case multiple counterbalance weights are necessary: they can be attached with Velcro, as is done by D. Parker. To improve the balancing of camera equipment mounted on Newtonian reflectors, and also to reduce the effects of vibrations, Gomez had the idea, about 20 years ago, of deflecting the image given by the telescope by means of a supplementary mirror set at 45°. The camera, even for high amplification (when far from the original focal plane) could thus be attached directly to the tube, its stability being increased and the imbalance reduced.

To enlarge the primary image of a Newtonian telescope, eyepieces are generally used. To calculate the amplification A of the image, the following formula given by Covington (1985) and Takahashi can be used: $A = P/f - 1$. Here P is the projection distance, or the distance between the focal plane of the eyepiece and the film, and f is the focal length of the eyepiece. The resultant focal length, F is given by $F = F_1 \times A$ (F_1 = focal length of telescope and A = amplification due to the eyepiece). It is easy to determine the new focal

---

* Observing visually a few minutes before and after Jean Dragesco's photographs of Mars and Jupiter, I can testify to their excellence: they record more detail than I could draw in large-scale sketches of individual features, and show *all* the details I could see through the eyepiece. The Mars photograph was taken during an exceptional period of perfect seeing, the extraordinary thing being that the planet, although near the meridian, was at an extremely low altitude at the time. This testifies, then, to the remarkable seeing conditions which can be had at the Pic. (RJM)

* For further details, see Thérin (1993).

ratio F/D. Unfortunately, the eyepiece is designed to give virtual images, i.e. images projected to infinity. As we wish to obtain a real image, at a finite projection distance, some spherical aberration will be introduced, becoming greater as the projection distance is reduced. Therefore, for a given magnification we must use rather powerful eyepieces, which will give the best images and reduce the encumbrance of the photographic apparatus (Miyazaki uses very powerful eyepieces down to 3.8 mm focal length). In other respects, eyepieces of average focal length are easier to use and sometimes of better quality. It is possible to overcome the spherical aberration introduced by image projection over short distances, by making use of a projection lens, i.e. a positive achromatic doublet, placed upon the eye-lens of the eyepiece; its focal length should equal the desired projection distance (and therefore the short-focus eyepiece is focused at infinity). Neřmec, who used a long-focus refractor ($F/D = 20$), could use simple eyepieces, Huyghenians or Kellners, without serious inconvenience. With modern telescopes of medium focal length, eyepieces such as Plössls, super-Plössls or Orthoscopics will do well, provided that they are of high quality (the Clavé Plössls are excellent, also the super-Plössls and the new 'Eudioscopics'). Their focal lengths should fall in the range 10–22 mm, preferably with long projection distances (12–18 cm). For Newtonians working at $F/D = 5–6$, the best possible eyepieces must be used, especially Näglers, of the short-focus kind. One can choose from a whole range of new high quality eyepieces produced by Vixen, Takahashi and Carl Zeiss Jena. Two Japanese manufacturers offer for sale eyepieces 'for projection' (calculated to project a real image over a short distance): Takahashi offer two, PJ-12 mm and PJ-20 mm; and Pentax offer four – XP-3.8 mm, MP-8 mm, XP-14 mm and XP-24 mm. These are the most interesting eyepieces for high resolution photography. The choice of a projection eyepiece for a short-focus Newtonian remains of fundamental importance.

Pope & Osypowski (1969), true pioneers of high resolution photography, proposed the use of *16 mm camera lenses* (of 12–17 mm focus), used the 'wrong way round' (with the front lens turned to face the film). Better still, Martinez (1987) advised the use of *objectives specially designed for macrophotography*, mounted upon a bellows system. That is what I had already used in 1980, at Cotonou, on my C14. The macro-objective of short focal length (35 mm) gives excellent images but requires precise optical alignment, and it

follows that a rigid mounting is essential. Rouse suggested using enlarging lenses (preferably turned backwards), but their focal lengths are much too long. That is why Dobbins *et al.* (1988) suggested the use of microscope objectives. This is an excellent idea in optical terms. An apochromatic, flat-field objective constitutes the best enlarging system possible and the projection distance can be quite short (16 cm). The use of a microscope objective could not be simpler: the objective should be screwed into a drawtube, with the frontal lens towards the focal plane of the telescope (the distance between the focal plane and the frontal lens, perhaps known in advance, being from only a few mm to 1 cm). The image furnished by the microscope objective should be used at a projection distance of 16 cm, then the magnification obtained will be that engraved upon the objective itself: ×10, for example, which will convert a $F/D = 6$ Newtonian into a $F/D = 60$ system. It is important to choose an objective with a flat field, even if it is rather more expensive. It is in the interest of the astrophotographer to use a numerical aperture as high as possible (0.30 or less, for a ×10 objective). It is possible to alter the projection distance slightly (16 cm in theory, but with low power objectives one can vary the distance by 3–4 cm, to obtain the desired focal length). For different magnifications, we must have access to several objectives: ×4, ×6, ×10, ×16, etc. The lowest power objectives need not be apochromatics. It is not easy to find the objective one needs; one must look through various manufacturers' catalogues before deciding what to buy (Leitz, Carl Zeiss, Nikon, Olympus). A good microscope objective is better than any eyepiece, but will not completely fill the standard format (24 × 36 mm) frame: this is, of course, unimportant for planetary photography. The ideal would be a zoom microscope objective, with magnifications from ×4 to ×12, for example. But such do not exist!

There is one further method of increasing the focal length of a given telescope: the use of one or more teleconverters. These are negative lenses which double the focal length of a photographic objective, with a projection distance practically zero. They correspond to ultra-compact, better-corrected Barlow doublets. Although Denis di Cicco asserted that the common teleconverters were preferable for astronomical use, I have always used either apochromatic teleconverters (Nikon TC-14 and TC-300) or good normal five-lens achromatic ones such as the TC-200 or the Foca 2 X. It is possible to combine teleconverter lenses to get a higher focal ratio. At the Pic du Midi and at Flagstaff I used either the TC-300 alone or combined with the TC-14, or two TC-2s joined together. It is worth pointing out that when we join up two high quality teleconverters, we end up having a system with up to 14 lenses (!), which is really too many. The main advantage of teleconverters is in their mode of use: easy to mount, compact, no optical alignment needed, but with low magnifications. Therefore, their use is limited. We would like to be able to use high quality teleconverters, of ×3 power or even ×4, but, again, they do not exist. We could even consider a zoom teleconverter giving powers from ×2 to ×5, but they would have to have 15–18 lenses, which would lead to considerable light loss and prohibitive expense!

The reader has possibly been swamped by the number of enlargement systems that I have described, so the essential details are summarised in Table 3.2.

3.2 Photographic equipment

The great majority of amateurs use 24×36 format reflex cameras for high resolution photography. The reflex system is an acceptable way of focusing and viewing, but with some reservations. Of course, except in the case of very heavy telescopes (the 520 mm of G. Viscardy or the 106 cm of Pic du Midi), practically all amateur instruments (even 400 mm instruments, weighing 500 kg) are very susceptible to vibrations introduced by the reflex action of the camera (when the mirror suddenly flips up out of the light path as the shutter is opened). Thus, the images obtained are blurred, with the resolution reduced to 3″.8 with a C8 and a Nikon! This is the big disadvantage of reflex cameras as such. However, cameras were formerly made where the reflex mirror could be smoothly raised with the help of a cable release: the Miranda Laborec, the special astrophoto Praktica, the Visoflex III body by Leitz (for the Leica M). There was also the Canon Pellix with fixed mirror. The shutters of these cameras were very smooth. It is occasionally possible to find the Leitz Visoflex III body (the model described as 'endoscopic', with interchangeable focusing screen, is the most interesting); also the Leica M1, Md, M2, M3, etc., with very smooth shutter release. Among the relatively modern reflex cameras, those which cause the least vibration seem to be the Olympus OM1, OM2 or OM3, which, sadly, are no longer

Table 3.2 *Efficiency of various systems for enlarging the primary image*

Type of telescope and focal ratio F/D	Enlarging system (resulting F/D = 40–120)						
	A	B	C	D	E	F	G
				(see below)			
Newtonian (F/D = 4–6)	–	–	–	–	**	***	–
Newtonians (F/D = 8–10), Schmidt–Cassegrains (F/D = 10) and refractors (F/D = 8–10)	–	**	–	*	**	**	**
Cassegrains, Maksutovs, Kutters and refractors (F/D = 15–20)	*	***	*	**	*	–	***
Cassegrains and Kutters (F/D = 25–30)	**	**	**	***	*	–	***

A : ×2 Barlow doublet.
B : Two ×2 Barlows joined together.
C : ×2 teleconverter.
D : Two ×2 teleconverters joined together.
E : Eyepieces (3.8–16 mm).
F : Microscope objectives (×6–×16).
G : Objectives for macrophotography, mounted in bellows.

made but can sometimes be found on the market. Note their low weight, smooth shutter release and Varimagni viewfinders; these cameras were well designed. On the other hand, Olympus makes the OM4-Ti, in titanium, an excellent camera but very pricy. A selection of equipment is illustrated in Fig. 3.1.

In astrophotography one will have difficulty in focusing upon and viewing a ground-glass or matt focusing screen, which is the sort normally supplied. If it cannot be removed (with models at the lower end of the price range), one must suppress the grain of the matt screen. This can be done by covering the focusing screen with a rectangular microscope slide, 24×32 mm, smeared with a little Canada Balsam. We can thus obtain a transparent (or clear) glass focusing screen, excellent for astrophotography but useless for ordinary photography. Happily, most modern camera bodies will allow the matt focusing screen to be swapped for another, transparent, one provided with a central graticule. This is the case with the Olympus OM4-Ti, the Nikon F2, F3 and F4, and the various Canon and Pentax models.

1

2

3

5

4

6

The focusing screen assembly (prism plus fixed-focus viewing lens) is not always particularly useful, as it entails viewing the image in the same direction as the camera (convenient with a Newtonian but not with a Cassegrain!). In the case of the Olympus, the prism being immovable, we must have recourse to an expedient: an additional viewfinder, called a Varimagni, is fixed upon the eye-lens of the OM to reflect the image once more (through 90°), to allow accurate focusing on the central graticule, at different magnifications. It is a good solution, though complex. Where possible, however, it is preferable to remove the whole of the focusing ensemble, and to replace it with a magnifying lens (in practice a high quality eyepiece), which enlarges the image 5–6 times, allowing the eye to focus precisely upon the focusing screen, with adjustment from −5 dioptres to +3 dioptres (Nikon F2, F3, F4; Canon F1; some Pentax). The aerial image obtained with these viewers (provided with clear focusing screens) is as good as possible (×5–6 is ideal for focal ratios F/D 30–60, but a little too much beyond that). For a focal ratio from 120 to 180, one must be able to swap the magnifying lens provided by the manufacturer for a simple doublet of 80 mm focus, mounted in a sliding tube, for focusing on the graticule, with a magnification of ×3 (to avoid an enormous, blurred, hard-to-focus image). To take care of any situation, the ideal would be a zoom-type magnifying lens, allowing magnifications from ×2.5 to ×8, with fully adjustable focusing.

In contrast, for small focal ratios of $F/D = 5$–10, I use a compound microscope for focusing, of high numerical aperture, magnifying ×40. It is most useful for precision focusing, when photographing the whole Sun or Moon.

Figure 3.1. Some photographic equipment used in high resolution astrophotography. *1. Nikon F2 camera with motor, mounted on sliding bellows for photomicrography ('ef' denotes the motor used for electric focusing). 2. Nikon F3 camera, with motor, right angle viewer and electrically operated shutter. 3. Nikon F2 camera with right angled magnifying viewer. 4. The same camera as No. 3, equipped with a microscope (m; × 40 magnification) designed by the author for ensuring precision focusing on the aerial image. 5. Robot Berning 24 × 24 mm camera (made in 1950). 6. Olympus OM2 camera, equipped with its Varimagni viewer, one of the most widely used reflex cameras by high resolution astrophotographers.*

When we order a transparent focusing screen, we must ensure that it is provided with a graticule (simple or double), to enable us to focus exactly on the screen.

It is useful to have a camera which can accept a film-dater (all the modern reflex cameras can do this perfectly) which imprints, upon the negative, the day and hour when each frame was taken. Unfortunately, hardly any of them will register, at the same time, the day, the hour, minutes and seconds. One cannot usually obtain more than the first three of these (with an uncertainty of 30 s on average, which is perhaps inconvenient for Jupiter, when one is thinking of making very precise positional measures on high resolution negatives). The film-dater records, usually: the day, the hour (in Universal Time) and the minute (corresponding to the start of the exposure). For example, 11.0.32 (11th day of the month, 0h 32m UT). One is therefore obliged to add by hand with a fine marker of the type Stabilo-OHPens, upon each strip of six negatives the year and the month (1992 June, for example). The inscription of these details on the film is a considerable simplification. Otherwise one must think in terms of using a dictaphone, and radio time signals! Above all, one must not make any mistakes regarding the number of the photos taken, if the essential details are to be written on each negative.

A motor allowing operation of the shutter and the automatic film advance can be a useful accessory which may save time. Systems for automatically measuring exposure time are rarely useful in astrophotography. One must therefore buy a camera which can work in 'manual' mode. However, automatic exposure can perhaps be used for the Sun or even for lunar photography, if the telescope is capable of absorbing the vibration caused by the exposure: that is to say, hardly ever!

It is hard to answer the question: 'Which is the best reflex camera for high resolution astrophotography?' One can answer that it is the Olympus OM4-Ti provided with a Varimagni viewer, with a clear (No. 12) focusing screen and a good motor (and not a 'winder', as it does not allow use of the 'B' time setting); or the Canon F1, with a right-angle viewer, clear focusing screen and motor drive; or, finally, one of the Pentax models, allowing the usual viewing screen to be swapped for a clear one, and equipped with a motor. I use either Nikon F2 camera bodies (purchased secondhand; not manufactured since 1970), entirely mechanically operated and very robust, which I have been able to complete with a DW-2 (×6) finder and a focusing screen of type M, or the Nikon F3 camera bodies, when the date needs to be imprinted, provided with DW-3 (×6) finder and motor drive. The Nikon F4 is a high performance camera body and more compact than the F3 with its motor, but its price is prohibitive. I also use the Leica M camera bodies (Mda, M4) with the famous Visoflex III 'endoscopic' attachment (with interchangeable focusing screens). The majority of modern cameras, entirely automatic (Nikon 801, Canon EOS, etc.), are of no interest for high resolution astronomical photography. We can sometimes find useful accessories for the old models on the market: right-angle viewers, clear focusing screens, motors, film-daters, etc.

Dobbins *et al.* (1988) insist upon the importance of vibrations caused by the release of the mirror in exposures with reflex cameras. To avoid this inconvenience, three solutions are open to us: (1) using the 'hat-trick' method, where a large external shutter covers and uncovers the aperture of the instrument; (2) installing a central shutter of the Compur type, in front of the reflex camera; (3) designing a device specially for high resolution photography.

Parker (1980) made a device which allowed him to monitor the image up to the last moment, the arrangement guaranteeing vibration-free exposure. The latest version of this system has been described by Dobbins *et al.* (1988): between the reflex camera body (24 × 36) and the drawtube, Parker introduced an auxiliary tube. The tube contains a flat mirror set at 45° which can be rotated about an axis (by means of an air-bulb cable release), thus allowing light to reach the film. The reflex camera is put on the 'B' setting and its shutter left open by a cable release. When the mirror is interposed, i.e. tilted at 45° (Figs. 3.2:1 and 3.2:2), the image of the planet observed is reflected into the lateral tube, provided with a 'seeing eyepiece', which enables the turbulence to be monitored. This eyepiece is chosen to be a long-focus one (80–100 mm), owing to the very high focal ratios which Parker likes. At $F/D = 160$, with his 400 mm telescope, he obtains a focal length of 64 m. A 50 mm focusing eyepiece will allow him to examine the image with a power of ×1280, which makes it blurred and faint. A 100 mm eyepiece allows him to focus with a power of ×640, which is very comfortable. A similar eyepiece not being commercially available, Parker made one for himself with the help of lenses from Jäeger (USA).

The mode of operation is very simple: the observer looks in the viewing eyepiece, focuses the telescope, centres his image, then monitors the seeing conditions (a graticule, situated at the focus of the eyepiece, allows for control of all movements of the image). After having opened the shutter of the reflex camera, Parker chooses the instants of best seeing for moving the small mirror, with the aid of an air-bulb cable release, and gives the exposure time he considers necessary (usually 3–8 s), with no vibration. By this procedure, we are also able to obtain several planetary images on the same frame (by slightly displacing the telescope in such a way that the images do not overlap). We can also use a film advance motor and a film-dater, but must then be content with a single image per negative.

In 1992 July I took up Parker's idea in the following manner. I use a Visoflex III Leitz camera attachment with focuser. It is provided with a clear focusing screen and I replaced its × 4 magnifier with a wide-field Meade eyepiece of 40 mm focal length (magnification × 6, ideal for focal ratios between $F/D = 40$ and 80). The aerial image is superb.* The reflex mirror of the Visoflex could be moved smoothly, thanks to a flexible cable release, but it was necessary to make a mask, keeping out all undesirable light which could pass around the mirror when the latter is lowered. To the Visoflex III reflex attachment, thus modified, is fixed a Leica M camera body provided with a flexible locking cable release. The mode of operation is simple: one observes the image in the eyepiece, which gives an aerial image of the planet to be photographed. The focus being made, the planet is placed at the centre of the field, tangential to one of the graticule lines. The shutter of the Leica camera is opened and held in that position with the aid of the locking cable release. The image is monitored, and, when the conditions seem to be optimal (smallest displacement of the image), one makes the exposure by smoothly raising the reflex mirror of the Visoflex. The mirror redescending, one closes the shutter of the Leica M and resets it (a Leitz mechanical motor can save time) (Figs. 3.2:3, 4, 5).

In 1992 I started to work again with an old camera which I had successfully used between

---

* During January 1993 I spent 2 months in Cotonou (Benin, West Africa) taking some 5200 astronomical photographs, especially with the new Leica–Visoflex combination. I have finally rejected the very heavy Meade UWF eyepiece as a magnifier and use the normal Leitz ×5 standard finder or a Peak ×10 magnifier which seems to be very good for focal ratios of 45–80.

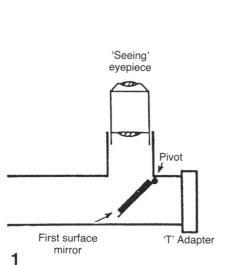
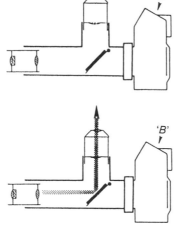
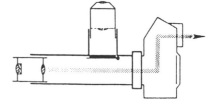
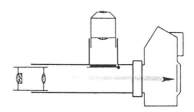

'Seeing' eyepiece

Pivot

First surface mirror

'T' Adapter

1

'B'

'B'

2

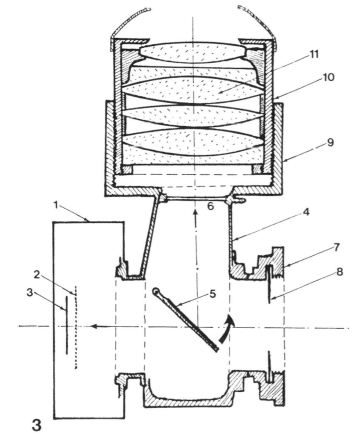

11

10

9

1

2

3

6

4

5

7

8

3

4

5

Figure 3.2. Parker's device for high resolution photography. *1. Pivoting mirror ensuring precise focusing and allowing image monitoring as well as serving as a manual occulter. 2. Schematic diagram showing the operation of the system (after Dobbins et al., 1988). 3. Use of the system by the author with a Leica M4 camera, equipped with a Visoflex III endoscopic viewer and a wide-field eyepiece. Key: 1, Leica body; 2, Leica shutter (on 'B' time setting);* *3, film; 4, Visoflex viewer; 5, movable mirror, serving as an occulter (by means of a cable release); 6, transparent viewing screen; 7, T ring provided with a rectangular diaphragm (8); 9, 10, 11, observing eyepiece (Meade S.W.A.). 4. Leica M4 camera with the Visoflex viewer, provided with its diaphragm ('d') (the latter is indispensable). 5. Lateral view of the set-up with the two cable releases (for a more detailed description, see pages 36, 37).*

1963 and 1971 (Bourge *et al.*, 1979). The camera is a 24 × 24 mm Robot Star of 1950 vintage(!). Small and light, this camera has a Compur-type shutter which is extremely smooth, allowing use of the 'B' time setting, and exposures between 0.5 and 1/500 s. The film is advanced by means of a spring-wound motor, allowing 24 exposures before rewinding is necessary. It is easily possible to take up to 60 exposures on a TP 2415 film, cut to a length of 1.7 m. The small camera body contains a fixed beam-splitter (a plane-parallel glass plate treated with a semireflective coating inclined at 45° to the optical train, which sends 30% of the light into the viewing eyepiece, whereas 70% is left to register on the film). This type of semireflecting mirror needs to have a delicate system for adjustment, in order to centre the image in the viewer with respect to the geometric centre of the 24 × 24 mm frame. The image is observed with the help of a Meade Plössl eyepiece of 40 mm focal length (magnification ×6) which allows the entire photographic field to be seen. A problem arises with the Robot 24 × 24: the narrowness of the shutter limits the width of

the quasi-parallel beam of light coming from the telescope (but not the beam leaving the camera lens, specially designed to be very close to the shutter opening), which reduces the useful format to only about 16 × 16 mm. I therefore used a negative doublet (a sort of powerful Barlow lens) designed by my friend Iorio Riva for microphotography with the Robot. This optical extra enlarges the image on the film, up to 22 × 22, which is adequate. The advantages of this outfit are severalfold: observation of the image quality is continuous, even during the exposure; it is also quick to use, allows a large number of photos to be taken on the same film, and is both light and compact. It has but two drawbacks: (1) the need to increase the exposure time by 30% (not very important for the Sun, but perhaps significant for the Moon and planets, because the exposure time may then exceed 2 s), and (2) the format is too small for lunar photography (but not a serious problem if we seek really high resolution) (Fig. 3.3). For one or other of these devices (again on trial), I expect to use microscope objectives, in order to make the device still more

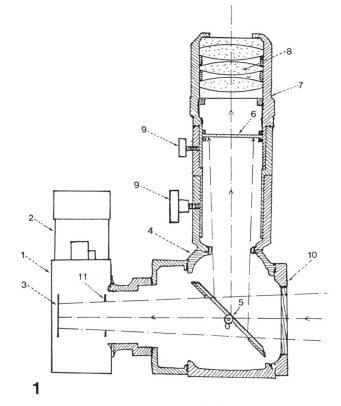

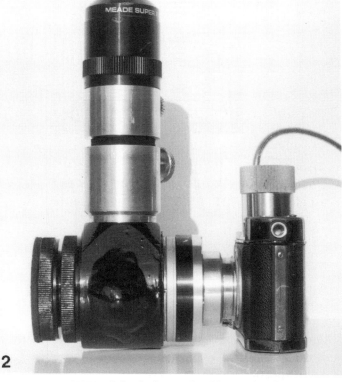

Figure 3.3. Astrophotographic device permitting continuous viewing of the image. *This arrangement is vibration-free. 1. Schematic cross-section showing: 1, Robot camera; 2, mechanical rewind; 3, film; 4, original reflex body constructed by the author, having a fixed mirror (5) tilted at 45° which* transmits 70% of the light to the film and reflects 30% into the viewing eyepiece; 6, transparent viewing screen; 7, 8, viewing eyepiece (Super-Plössl, 40 mm); 9, clamping screw; 10, T ring; 11, Compur shutter of the Robot camera. 2. General view, as published recently by the author (Dragesco, 1993).

compact. The Robot camera was designed in 1934 by H. Kilfit, and has been made since then by Otto Berning. A well-designed model is the Robot made for the Luftwaffe, which is provided with a large motor-winder allowing it to take up to 50 photographs without rewinding. One can very easily find the Robot I, Robot II or Robot-Star at affordable prices (on the other hand, the special Robot cartridges are very difficult to find). In fact, the firm of Robot Berning was taken over in 1964 by Pierre Couffin. The Robot is therefore still being made, but its structure is now completely different. It is made more for its industrial and scientific applications, and its current price is unfortunately very high. For planetary photography, which requires a small format, one can seriously think of using a half-frame (18 × 24 mm) camera, or even a small 35 mm cine-camera. With a small telescope 16 mm film could also be used (with an available image size of 7.5 × 10 mm); but again the astrophotographer would have to be able to find TP 2415 film in this size.

3.3   Taking high resolution views

Following from the above, it is not advisable to use an ordinary reflex camera in the normal way, because of the vibrations which result from the action of the reflex mirror. I looked into the degree of blurring introduced by these vibrations: the Nikon F3, used for 1 s exposures, reduced the resolving power to an average of 3″.8 on a Celestron 8, potentially capable of resolving to 0″.6. Even an exposure time of 1/30 s was harmful, and the resolving power was far from the theoretical limit (it is limited to about 2″). However, this is an exposure time often used by amateurs: the whole Moon, solar eclipses, etc. The results will be better with an Olympus OM. For my photos of the solar chromosphere (in H-alpha), I had to have recourse to a Leica M camera body, which allowed me to manually (and thus very gently) raise the mirror of the Visoflex III, with the help of a cable release. On the other hand, the damped shutter blinds cushion the action of this camera, assuring a resolution of about 1″, with 1/30 s exposure. With a normal reflex camera, any exposure time slower than 1/250 s allows a resolution of 1″, good enough for the Sun (but it is always better to use speeds of 1/500, 1/1000 or 1/2000 s).

To avoid the vibrations due to the reflex action, we can usually use the so-called 'hat-trick' method, i.e. we use a large board, which serves as

a shutter for the telescope tube itself. But this procedure is slow and painful. The operation goes like this: we focus on the outline of the planet in the centre of the clear glass viewing screen of the camera, and then use the right arm to cover the open end of the telescope tube with a cardboard disk. The disk should be 5–6 cm wider than the aperture, and can be made of thin plywood or plastic (etc.) painted matt black. With the left hand we open the shutter of the camera, on the 'T' or 'B' time setting (if we possess a locking cable release). Then, with the right hand, we hold the large disk about 5–6 cm in front of the telescope and wait several seconds for the vibrations to die away; then we remove it, to the side, in a sudden sweep, thereby uncovering the telescope aperture for the necessary time (1–3 s). Finally, we cover the telescope with the large disk and, with the left hand, close the camera shutter. If a motor is available, one can immediately begin a second cycle of operation; otherwise one must reset the camera manually. With a 200–250 mm Newtonian, it is a relatively easy operation. It is already becoming tiring with a 350–400 mm Cassegrain, and it becomes extremely painful with a 150–180 mm refractor (the large disk must then be mounted upon a long wooden pole!). To make the disk, one should choose a light but rigid, flat material: thin Plexiglas, 3-mm-thick plywood, a sheet of thin aluminium, etc. The two faces should be painted matt black, but one should provide a thin ring of white paint around the circumference of the disk so as to be able to align the disk in the dark. One must provide a handle, again a very light one, to enable the disk to be moved: balsa wood is excellent.* This widely used procedure has three drawbacks. For short exposure times (0.5– 1 s) the mere act of uncovering and then recovering the objective, during part of the exposure, involves an alteration of the diffraction pattern: the resolution is then not the same from (say) left to right as from top to bottom, depending on the direction of movement of the external 'shutter'. In other respects, the manipulation of the disk, in front of the open tube, risks upsetting the air layers in the vicinity of the aperture. We work, therefore, 'in the dark', without the choice of taking the photo in the rare moments of good seeing. We can imagine a small, very light shutter, placed in front of the camera, being

---

* C. Arsidi and G. Thérin take the trouble to put a glove on the hand which holds the full-aperture shutter. This ensures that no heat from the hand can introduce turbulence into the light-path (Thérin, 1993).

swung aside by a cable release. Some colleagues have installed a central Compur-type shutter at the entrance of the reflex camera body. Comfortably installed, the observer now views the focusing screen, then closes the Compur, opens the shutter of the reflex camera ('B' setting), waits four or five seconds and then makes the actual exposure, with the aid of the Compur shutter. One now has only to close the reflex camera shutter and to begin again the sequence of operations (a rewind motor makes the job more pleasant!). First, one must have tested the Compur shutter, to ensure that it cannot introduce any awkward vibrations. For that, the Compur must be attached to the drawtube (with adhesive tape), and a star then observed with a higher power (×300). One sets and releases the shutter several times (above all, on the 'B' setting) and checks for the total absence of all vibrations in the image of the star. The inconvenience of this system resides in the impossibility of observing the image just at the moment of the shutter release (again working in the dark). Sometimes, also, small vibrations may remain in the use of a Compur shutter, which do not allow the maximum resolving power to be reached.

The best system, therefore, seems to be that adopted by Parker or the modification that I have described more fully (Fig. 3.2). I also think that the device provided with a Robot 24 × 24 camera and a viewer for the continual monitoring of the aerial image could be an interesting combination. A detailed description of this type of apparatus has been given elsewhere (Dragesco, 1993).

3.4 Exposure times in high resolution photography

Theoretically, exposure times in high resolution photography depend upon a large number of factors: the albedo of the object concerned, overall focal ratio F/D, sensitivity of the film used (which varies considerably according to the type of developer and development time), density of any filter used, etc. We cannot simply be content to work out the exposure time and leave it at that. We often have to vary the different parameters, to obtain an exposure time which could result in the best possible photographic image.

The brightness of each planet varies over very wide limits. After Dobbins et al. (1988), the brightness ratios and the theoretical exposure times vary according to the data in Table 3.3 (slightly modified and extended).

On the question of image brightness, we know

Table 3.3 *Brightness of the Moon and planets, and relative and actual exposure times*

Object	Relative brightness	Exposure times (s) relative (in theory)	in practice
Venus	75 000	1/20	1 [a]
Mars	4 250	1	1–2 [b]
Jupiter	1 150	3.5	2–3 [c]
Saturn	350	10	4–5 [d]
Moon (first or last quarter)	1 100	3	1–1.5 [e]

Corrections to the theoretical exposures (for F/D = 50–80 approx.)
[a] Increase the focal ratio and add a filter to enable exposure for 1 s (at 1/20 s the image will be blurred because of the vibrations introduced by the shutter of the reflex camera).
[b] Exposure times of 1–2 s correspond to focal ratios of 60–80, with TP 2415 film, using a Wratten 25 red filter.
[c] For Jupiter, one must try to reach a focal ratio of $F/D = 80$–90 without filter or of only $F/D = 60$ with a blue filter (6–8 s exposure).
[d] With Saturn, one must not be tempted to exceed 5 s if one wants good resolution. Complex problem: the planet is small and one cannot reduce the focal ratio too much ($F/D = 60?$). Saturn remains a difficult object for small instruments.
[e] The Moon requires short exposures, because of the possible significant speed variation in RA and declination. We can be content with $F/D = 50$–70.

that the exposure time varies with the square of the focal ratio F/D. For example, if at $F/D = 10$ the exposure is 0.1 s, at $F/D = 70$ it will exceed 5 s ($7 \times 7 = 49$). This is summarised in detail in Table 3.4 (Dobbins et al., 1988, completed by the author).

In addition to what has been stated above, there are two schools of thought in high resolution astrophotography: (1) obtain large images on the negative ($F/D > 120$) and so increase the exposure time (3–12 s); (2) reduce the focal ratio ($F/D = 40$–80) in order to enable the exposure time to be reduced (ideally to between 1 s and 2 s). The character of the turbulence at a given site and the quality of the telescope drive play a decisive role in the choice of one or other of the above solutions. The exposure time, essentially, depends upon the brightness of the celestial object in question, the focal ratio, the sensitivity of the film (100–400 ISO), the transparency of the sky (whether dry or humid, cloudy – and therefore upon the elevation of the site) and the altitude of the object above the horizon. For a given film we must also take into account the exposure factor of any filter used (which, again, will increase the exposure).

Table 3.4 *Increase in exposure time as a function of focal ratio (modified from Dobbins* et al., *1988)*

F/D	5	10	20	30	40	50	60	70	80	100	120	150	200
Exposure time [a]	0.01	0.04	0.16	0.32	0.65	1 [b]	1.4 [b]	1.8 [b]	2.6 [b]	4 [b]	5.6	8	16

[a] Average actual exposure times with TP 2415 film, developed under forcing conditions.

[b] The best exposure times, in my personal opinion.

3.5 The role of filters in high resolution photography

With a small instrument it is always preferable to take photographs without a filter, in order to shorten the exposure time and thus to obtain higher resolution. However, the use of filters is desirable, as, thanks to them, we can enhance various types of planetary features (clouds on Mars in blue light, intensification of the plumes in Jupiter's equatorial region in orange light, etc.) or photograph others which are otherwise invisible (such as the UV markings in the Venusian clouds). It is also possible, thanks to the use of monochromatic filters (dark green in preference), to improve the quality of a refractor's objective when there is a noticeable secondary spectrum. Finally, we are obliged to use colour filters (well-chosen ones) to obtain high resolution colour photographs by the trichrome process (employing panchromatic TP 2415 film, used successively with three complementary filters, followed by compositing in the darkroom).

The filters most often used are Kodak Wratten gelatine filters (Eastman Kodak, 1981) which are currently sold in three sizes: 52 × 52 mm, 75 × 75 mm and 103 × 103 mm (one can also get 150 × 150 mm and 300 × 300 mm on special demand).

Wratten 96 filters (see Tables 3.5 and 3.6) are grey or neutral filters of variable density, which allow us to vary the amount of light at will. Those with transmission 1/1000 and 1/10 000 can sometimes be useful in solar photography, provided that they are used in front of the objective. But Wratten filters are very fragile: they are easily scratched, warp in strong sunlight and are damaged by moisture. One must keep them in a dry box, free from moisture and dust.

For *trichrome* work, using three Wratten filters, Wallis & Provin (1988) advise the use of the W47, W57A and W23A.

For *UV* work (Venusian clouds), Martinez (1987) recommends the Schott UG5 glass filter (which was also used by C. Boyer and by myself).

Table 3.5 *Density and transmission characteristics of Wratten 96 neutral density filters (reproduced by kind permission of Société Kodak-Pathé)*

Density (log)	Transmission (%)
0.10	80
0.20	63
0.30	50
0.40	40
0.50	32
0.60	25
0.70	20
0.80	16
0.90	13
1.00	10
2.00	1 (1/100)
3.00	0.10 (1/1000)
4.00	0.01 (1/10 000)

Table 3.6 *The main Wratten colour filters for astrophotography (reproduced by kind permission of Société Kodak-Pathé)*

No.	Colour	No.	Colour
3	Light yellow	47	Medium blue
4	Medium yellow	47B	Deep blue
8	Stronger yellow	49B	Dark blue
9	Intense yellow	50	Monochromatic blue
12	Very deep yellow	36	Dark violet
21	Orange	58	Medium green
23A	Light red	61	Deep green
25	Medium red	74	Monochromatic green
29	Deep red	87	(Transmits infrared)
92	Very deep red	18A	(Glass) + W36 (transmits ultraviolet from 300 nm to 400 nm)

Rejection filters For solar photography we can use rejection filters, which eliminate the excess light and heat emitted by the Sun. For photography in white light we must reject, by reflection, up to 999/1000 (99.9%) of the solar energy. It is often possible to use Mylar film (aluminised thin plastic sheeting), which transmits about 1/1000, or 0.1%, of the solar energy. In theory, filters must alter the diffraction pattern, but as atmospheric turbulence has a much greater influence, the images obtained are very satisfactory for a typical diameter of 10 cm (resolving power about 1".2). I think that when the objective and seeing are both good, it is always preferable to use rejection filters of plane-parallel optical glass (such aluminised 'windows' will not change the image quality) rather than Mylar film. These filters require high quality optical glass, delicate working and a high precision polish; therefore, they are necessarily very costly. One must be suspicious of the cheap glass filters sold by certain American firms. On the other hand, the filters sold by the Schmidt–Cassegrain (Celestron, Meade) or Maksutov (Questar) telescope manufacturers are of the necessary quality but often too opaque (transmission between 1/10 000 and 1/100 000), and therefore unfavourable for high resolution photography which needs very short exposures of the order of 1/1000 s. These filters were designed for visual observation, following the normal stringent safety requirements. It is possible to obtain good aluminised glass rejection filters from some well-known firms such as Carl Zeiss Jena and Lichtenknecker (for more details see Section 4.2.3, pp. 67–8).

For Hα photography (λ = 6569Å: 656.9 nm, we use red glass rejection filters, of relatively low absorbance. They reject 80% of the total energy (mostly from the infrared region), remaining very transparent in the Hα region. For more details about these filters, see pp. 79–80.

The use of filters is very important in high resolution photography. The most practical are still the Wratten gelatine filters, but they are fragile, being very sensitive to strong humidity: thus just a few hours under a humidity of 95% is enough to completely ruin a Wratten filter. In view of their relatively modest price, it is easy to replace damaged ones. It is best to buy a larger-format sheet (102 × 102 mm), from which can be cut several small circular filters. Glass colour filters may also be used: they are much more resilient and can be obtained from Lumicon, Schott, Corning, MTO, etc. The Belgian firm Lichtenknecker produces a series of glass filters: GG 495 (yellow), OG 550 (orange), RG 610 (red), RG 715 (very deep red), UG1 (for ultraviolet work at 355 nm), BG 14 (blue) and VG6 (green). Carl Zeiss Jena can provide a whole range of neutral filters (t = 0.007–0.70) as well as colour filters: GG7 (yellow), OG5 (orange), RG2 (red), VG9 (green) and BG12 (blue). The transmission curves

of these filters can be obtained by writing to the manufacturer. Celestron and Meade also make coloured glass filters on demand, as well as polarising filters, which can be used like attenuators: turning one with respect to a second polarising filter allows a large range of transmissions to be obtained. Finally, Schott has a very wide range of high quality glass filters.

3.6 Photographic considerations

3.6.1 Elementary sensitometry

It will be useful to recall here some ideas on the fundamentals of emulsion sensitivity. For more details see Glafkides (1976), Eastman Kodak (1977), Eggleston (1984), Todd and Zakia (1986), etc.

As de Vaucouleurs et al. (1956) state, the coatings of photographs are gelatine–silver bromide emulsions, containing a microcrystalline suspension of silver bromide and iodide, in a gelatine gel. The crystals are of various forms (derived from the cubic system), and their dimensions vary with the sensitivity of the emulsions: 0.1–0.05 μm for slow emulsions and 1–2 μm for very rapid emulsions. In fact there is always a mixture of grains of various sizes, particularly in those emulsions said to have a wide 'exposure latitude'. A fresh, sensitive coating will not show any visible change after exposure to light. The image is therefore said to be 'latent', as it has to be revealed by the action of a developer. Once the sensitive emulsion has been developed and fixed, the crystalline silver halides will have been reduced to opaque, metallic silver. The darkening of a developed photographic image is characterised by its optical density (in diffuse or direct light). Let F_0 be the incident luminous flux (or light intensity), F the transmitted luminous flux (after passing through the darkened emulsion), T the transmission factor and O the opacity (sometimes known as absorbance). Then we can write $D = \log F_0/F = \log O = \log (1/T)$. That is to say that the density D varies between about 0 and 3 for the maximum opacities of 1000 and the minimum transmissions of 1/1000. To determine the properties of a photographic emulsion, the density of the darkening is measured for different degrees of illumination. By plotting the density D against the logarithm of the illumination, we obtain a curve known as the characteristic curve, or the curve of Hürter and Driffield. To vary the illumination, one can either vary the exposure time, having chosen a constant brightness, or vary the brightness and keep the exposure the same, as the illumination is the product of the exposure E and the time t. Examination of this exposure curve (Fig. 3.4:1) reveals three main regions: (1) the region of the concave curve AB, where the film has been underexposed; (2) a nearly linear portion BC, called the normal exposure region, and (3) a convex curve portion CD, which represents overexposure. It is only in the middle region BC that there is a reciprocity (or inverse relationship) between the illumination and the image density: over the exposure range 0.001–1 s, approximately. That is, we obtain the same darkening by varying E and t in an inverse sense, remembering that $E \times t$ = constant. This is known as the *reciprocity law* (due to Bunsen and Roscoe).

The slope of the linear part of the curve is known as the *gamma or contrast factor* (γ). Thus, the tangent to the linear part of the curve determines the contrast of the negative. Its value is logarithmic and varies between about 0.3 and 4. When the slope of the tangent is 45°, γ is unity and we say that the contrast is normal, as the darkening is exactly proportional to the illumination. A γ of 0.6 therefore represents a lower contrast than normal (useful in landscape or portrait photography, as, during enlargement onto paper, the contrast will be increased). On the other hand, a γ between 2 and 3 corresponds to very contrasty negatives, useful in scientific photography, allowing for the recording of details that are normally hardly visible. For several years now, the idea of a gamma factor has been

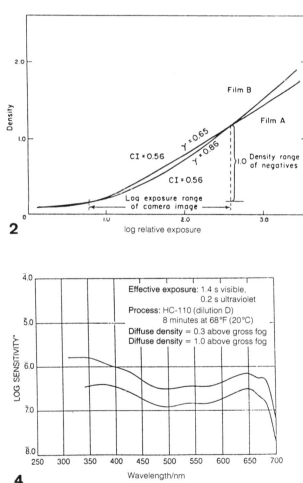

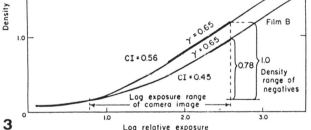

Figure 3.4. Characteristic curve *(or the curve of Hürter and Driffield), where AB is the region of underexposure; BC is the region or normal exposure (linear part); CD is the region of overexposure; V is the density of the fog; and gamma is the contrast factor, after de Vaucouleurs et al. (1956). 2. Two films A and B can show different gamma values but* have identical Contrast Indices. 3. Two other films A and B can have the same gamma value but different Contrast Indices (for more details see page 44). 4. Spectral sensitivity of Kodak TP 2415 film; notice the high sensitivity in the red (650 nm) and the low sensitivity in the blue (480 nm). Parts 2, 3 and 4 courtesy Eastman-Kodak.

abandoned, and it has been replaced by a Contrast Index (CI), which presents values a little lower than the gamma factors. Eastman-Kodak (1972) believes that the idea of γ is valuable for films with characteristic curves having a long linear portion. Numerous modern films have curves virtually devoid of a linear part. It is now preferable to include the 'foot' of the curve (i.e. the underexposed region) in forming an idea of the contrast, to define the so-called CI: 'Two films developed to equal gamma can have very different toe slopes. The Contrast Index is the slope of the straight line drawn between the two points of the $D/\log E$ (characteristic) curve that represents the desired minimum and maximum densities of high quality negatives.' The minimum density is an arc of a circle of radius $D = 0.2$ and the maximum density is an arc of a circle of radius $D = 2.0$. The two circles are concentric and their centre passes through the origin of the graph (i.e. the density of the unexposed sensitive emulsion, beneath the density of the base) (Figs. 3.4:2, 3).

In practice, in high resolution photography we use contrasty, well-developed films; the characteristic curves obtained lend themselves more to measures of γ than to the CI. In the remainder of this book, I will sometimes use γ values and sometimes CI values, according to the documents in my possession (the CI values always being lower).

The problem of obtaining contrast, an essential quantity in high resolution photography, has given rise to widely differing views, sometimes nonsensical ones. Dobbins *et al.* (1988) consider that planetary photography has the same requirements as photography of everyday subjects: landscapes, for instance. Under the conditions necessary to permit of photographing all the details of an image, one must avoid having a high contrast if the film is to be capable of recording the numerous light levels which we see in nature. This is exactly the case for so-called 'pictorial' photography, where one must obtain moderate contrast ($\gamma = 0.7$ or CI = 0.5). That is why American writers recommend a CI of only 0.5 for the Moon, 0.6 for Mars and 0.7–0.8 for Jupiter. I am in total disagreement with this! In a sunlit landscape one can find extreme variations in brightness, with brightness ratios reaching 375:1, whereas the best photograph, printed upon brilliant white paper, cannot record a range in brightness of more than about 50:1 (Adams, 1948). In the case of planetary surfaces, the observable details show small variations of brightness levels; the contrasts remain moderate

and are further decreased by the effects of diffraction when we are working at the limit of resolution of the telescope. Therefore, we must use a technique that increases contrast. Besides, it is easy to try the following experiment: if we photograph Jupiter, with Kodak T max 100 film and develop it in a special T max developer to a CI of 0.8, we will obtain a completely grey, and therefore useless, picture. As Martinez (1987) wrote: 'in planetary photography we must obtain a contrast index higher than unity'. Viscardy (1987) goes still further: he considers that, even for lunar photography, an extreme contrast is necessary if the details at the limit of resolution are to be recorded on the negative. Moreover, for his famous atlas of the Moon he used images where γ reached 3.20 or more! Here it is perhaps a question of exaggeration in the opposite sense to that of Dobbins *et al.* (1988). At $\gamma = 3.20$ a lunar photograph, obtained in good seeing, can show details at the limit of resolution of the telescope, but the negative will be extremely difficult to print and may require an unsharp mask for so doing. In reality, extremes should be avoided. If we look at the working conditions of the high resolution photographers, we perceive that the contrast of their photographs varies within certain limits. For the Moon, there is a tendency to want to obtain average contrast, which readers it is very easy to obtain very beautiful enlargements on paper, with numerous grey levels, brilliant whites and intense black shadow details. However, if the photographer seriously wishes to attain high resolution, it is necessary to obtain contrasts as high as $\gamma = 2.30$. For Mars, $\gamma = 2$ gives good results, but Jupiter may need higher contrast ($\gamma = 2.5$ or CI = 2.0). Saturn demands to be treated differently, depending on whether it is a question of recording the global details or the ring details ($\gamma = 2.0$–2.5). In a general way, in high resolution photography we are always seeking a relatively high contrast, always higher than unity ($\gamma = 1.3$–2.8; CI = 1–2). That is why solar, lunar and planetary photography is hard to achieve with normal (or commercial) black and white or colour films.

3.6.2 *Photographic emulsions for high resolution photography*

In the past, all astronomical photographs were taken on glass plates, in large format (4×5 in, 7.5×10 in), including the lunar, solar and planetary photographs of the first half of the twentieth century (1900–1950). Some observatories still use glass plates, as one can find a great variety of sensitive, specialist emulsions (Kodak Spectroscopic plates, for example). One reason for the success of the lunar and planetary photographs obtained at Pic du Midi from 1942 to 1950 was the use of fine-grain Agfa plates of relatively high contrast. When the photographic quality of most ancient pictures leaves something to be desired, it is because there was no sensitive emulsion that united several often contradictory characteristics: great sensitivity, high contrast, high resolving power and ultra-fine grain. The amateurs were even worse off, as they had to work with 35 mm films, which were poorly suited to high resolution photography. The pioneers of that time, such as G. Viscardy, G. Nemec, H. E. Dall, etc., recall the difficulties they encountered in finding the most suitable films. Some preferred the fastest (but very grainy) films, even though the image had to be greatly enlarged ($F/D = 200$). These films were lacking in contrast, even when developed (or forced) with energetic developers (D19). Others, wiser perhaps, preferred to use rather slower, but finer-grained and more contrasty, films. However, when the focal ratio F/D fell to 20 or 30, the planetary images were tiny. G. Nemec obtained good results with a film which was much in demand in the period 1955–65: Adox KB14 (14°/10 DIN) which represented a good compromise, having fine grain and yet adequate sensitivity. Others chose between Adox KB17, Kodak Panatomic X, Ilford FP3 or Agfa Isopan F. For the 106 cm telescope at Pic du Midi, Camichel and Boyer, the resident experts in planetary photography, found it advantageous to use Ilford Pan F, the sensitivity of which remained rather low (50–150° ISO), but the grain and contrast of which were acceptable (when developed in D19). I used it with some success (Fig. 4.46:5) in 1975. It was on Pan F film that C. Boyer succeeded in taking, on 1975 October 2, a photograph of Jupiter that remained for the next 12 years the best in the world (with a resolution less than 0″.20).

Then came the film which we all agreed in calling 'the miracle film': Kodak Technical Pan 2415 (which had been preceded by Technical Pan SO-511, specially formulated for solar photography and photomicrography). TP 2415 is quite a unique film, the only one which successfully combines very high resolving power with a good general sensitivity and a high contrast. It is, moreover, an extremely versatile film, the characteristics of which can be modified at will, depending on the development technique.

TP 2415 has therefore taken over as the universal film for amateur astrophotography. Used from the start, particularly for high resolution solar and planetary work, it also gradually became the ideal film for deep-sky photography, as it responds well to hypersensitisation by forming gas. The support of TP 2415 is 'Estar', which has excellent dimensional stability, making it highly suited to astrometry.

3.6.3   A special study of TP 2415 film

Spectral sensitivity   This film can be used from the violet up to the infrared. The curves published by Kodak (Fig. 3.4:4) do not agree very well with the results obtained by amateur astrophotographers (Dobbins *et al.*, 1988; Dragesco, 1988a,b). If the sensitivity in the ultraviolet is satisfactory, the blue sensitivity is disappointing, the use of the Wratten 47 B filter requiring very long exposure times (Table 3.7). The sensitivity in green light is very poor; it is much improved in yellow light, and becomes excellent in red, even up to 680 nm (for H-alpha solar work). In fact, the only small fault which can be found with the film is its low sensitivity in blue light. Planetary photography in blue light is important, so one is obliged to reduce considerably the focal ratio F/D, as using any blue filter with TP 2415 increases the exposure time significantly.

Overall sensitivity   According to Kodak, TP 2415 should be a film of average sensitivity. In fact, its sensitivity can vary over the enormous range of 1–12, according to development technique: 20°

Table 3.7 *Exposure factors for the main Wratten colour filters used with TP 2415 film (Dragesco 1988a,b)*

Filter colour	No.	Exposure factor[a]
Deep violet	W 66	5×
Deep blue	W 47 B	9×
Medium blue	W 47	8×
Blue-green	W 59	5×
Medium green	W 58	8×
Yellow	W 8	1.2×
Orange	W 21	1.5×
Light red	W 23 A	2×
Deep red	W 29	4×
Red hydrogen alpha	Lumicon hydrogen alpha (glass)	5×

[a] These factors were all obtained with hypersensitised TP 2415 film.

Table 3.8 *Variation of the overall sensitivity of TP 2415, according to the method of development (compiled from data after Dobbins et al., 1988; Kodak; Dragesco, 1992)*

Type of development (all at 20 °C)	Technidol 15 min	D76 12 min	Rodinal (1:50) 14 min	HC-110(B) 12 min	D19b 6 min	Dektol 3 min
Sensitivity (° ISO)	25–30	125	150	250	150	200

Table 3.9 *Contrast obtained with TP 2415 film, depending on the method of development (from data by Kodak and Dragesco (1992), and recent personal research; we thank Société Kodak-Pathé for their courtesy; see also Fig. 3.5).*

Type of development (all at 20 °C)	Technidol 15 min	D76 12 min	Rodinal (Agfa)[a] 14 min	Dektol 3 min	HC-110 (dilution B) 12 min	D19 or D19b 6 min
CI	0.8	2.10	1.00–1.60	2.50	2.10	2.35
γ	1.0	2.60	1.25–1.70 following agitation	3.60	2.60	3.20

[a] With Rodinol (Agfa) developer, one can obtain a large range of contrast according to the dilution, the mode of agitation and the development time.

ISO from underdevelopment in Technidol, 150° ISO after treatment with D19b, and up to 250° ISO when overdeveloped (or forced) in HC-110. This last value is remarkable and especially interesting for lunar and planetary photography. See Table 3.8.

Contrast By definition, TP 2415 is a film predicted to provide high contrast (Table 3.9). However, Kodak have perfected special developers for it, giving the possibility of using this film in the field of pictorial photography (countryside, industrial photography) with an exceptionally high resolving power. The developers Kodak Pota and Kodak Technidol allow the contrast index of TP 2415 to fall to 0.5, but the overall sensitivity is then no higher than about 25° ISO. This technique is of no interest in high resolution photography, which requires higher contrasts than CI = 1.2, and the highest possible speed. However, the treatment with Technidol (overdevelopment by at least 30%) is perhaps interesting in whole-disk lunar photography around First or Last Quarter, for it can reveal incredibly fine detail. The reduced sensitivity is in this case an advantage, as one can use an exposure of 0.5–1 s if necessary (perhaps by means of the 'hat-trick'), and also add a green filter. For Full Moon, the film must be more vigorously developed (Rodinal at 1:50 dilution) and a neutral density filter used in order to enable work to be carried out in the 0.5–1 s exposure range (which avoids using the reflex camera shutter). However, this will not be high resolution photography!

In high resolution photography we are looking for a high to very high contrast, which will vary somewhat with the object in question: $\gamma = 1.5$–2 for the Moon, 2–3 for the Sun and planets. A γ of 4 can perhaps be sought in extreme cases, such as photography of the solar granulation. The CI will vary, under the same conditions, between 1.2 and 3.

The *resolving power* of TP 2415 is very high but varies with the type of development used: more than 300 lines/mm for contrasty subjects after development in Technidol, but noticeably fewer for less contrasty subjects (80–130 lines/mm, according to the type of development). The study of modulation transfer curves shows that there is a loss of contrast as soon as the spatial frequency falls below 50 cycles/mm. The result is worse when an 'active' developer and overlong development times are used, the latter resulting in an increase in the granulation of the emulsion.

According to Kodak, the resolving power of TP 2415 varies between 300 and 400 lines/mm for the highest contrasts (1000 : 1) as a function of development technique, and between 100 and 125 lines/mm for the lowest contrasts (1.6 : 1). According to Dobbins *et al.*, 1988, the resolving power of TP 2415 can fall to 80 lines/mm, after overdevelopment (forcing) in concentrated HC-110. Be that as it may, the resolving power remains very high and is thus especially good for high resolution astrophotography, even at $F/D = 40$–50.

The *grain* of TP 2415 is extremely fine. Even when overdeveloped in the very active developers such as D19 and Dektol, the grain remains fine and allows for significant enlargements (up to ×14). The sensitive layer is very thin; it has good acutance and is little affected by irradiation or the halo effect (thanks to an efficient anti-halo layer).

Having a thin emulsion leads to an additional advantage in that fixing and washing require little time. The film keeps well, even at 20–22 °C. It is not necessary to keep it refrigerated, except in hot countries or for extended periods of time.

The *reciprocity failure* of 2415 is significantly high. Recalling that we spoke of the reciprocity law earlier (de Vaucouleurs *et al.* (1956)), since $E \times T$ is a constant (the law of Bunsen and Roscoe), we can obtain the same degree of

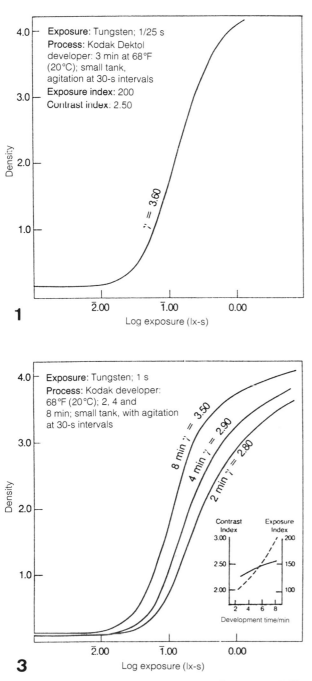

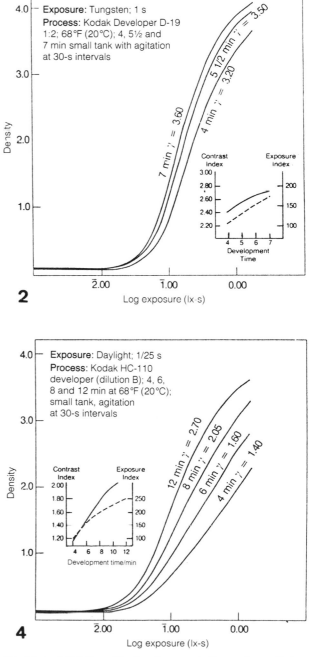

Figure 3.5. Characteristic curve of TP 2415 film, *after treatment with various developers. 1. Kodak Dektol (gamma = 3.60!). 2. D19 diluted 1:2 and with development times between 4 and 7 min.*

3. Undiluted D19, with development times from 2 to 8 min. 4. Kodak HC-110 (dilution B for 4–12 min). Courtesy Eastman-Kodak.

darkening of the emulsion by varying E and T in an inverse ratio. The law of reciprocity is applicable for the majority of sensitive emulsions over the exposure range 1/500–1 s. For exposures less than 1/1000 s or greater than 3–4 s, we observe a breakdown in the reciprocity law, or reciprocity failure (the Schwarzschild effect). Each emulsion reacts in a different manner and thus presents a reciprocity failure unique to that emulsion. TP 2415 film obeys the reciprocity law relatively well up to an exposure time of about 4 s. For very long exposures (20 min or more), the reciprocity failure becomes considerable. Happily, this film reacts incredibly well to gas-hypering (Dragesco, 1987a,b). The optimal duration for the hypering treatment is between 38 and 48 h, under a pressure of 3 psi (or 0.2 bar, etc.) at 55 °C; the hypering time exceeds 80 or 90 h at 50 °C; and at 60 °C it falls to 26 h. Gas-hypering (or hypersensitising) is of more interest in deep-sky photography but can also give useful service in high resolution work. A hypered TP 2415 film has the following advantages: a small gain in sensitivity (30% for 2 s exposures and 100% for exposures of 8 s); a slight reduction in contrast (useful for lunar work when one must develop fully in order to make use of the sensitivity of the emulsion); and an improvement in the foot of the characteristic curve, catching a greater degree of detail in dark regions (Reynolds & Parker, 1988; Dragesco, 1988a,b). Hypered film must be kept in the ice-making compartment of a refrigerator, or in a freezer, at −25 °C. Alternatively, it can be stored in a closed box containing a desiccant, either under vacuum or in forming gas. For the best hypering techniques, Heudier *et al.* (1981a,b), the literature by Lumicon or the publications of the author (Dragesco, 1987,a–c; 1988a,b) can be consulted.

TP 2415 film is particularly useful in 35 mm format (135–36 exposures, code 1297563) or in 45 m reels (codes 651 and 1299916). Larger formats also exist: 120 roll film size (6 × 6, 6 × 7 or 6 × 9 cm, codes 6415 and 151.1054); or boxes of 25 flat sheets of 10.2 × 12.7 cm (4 × 5 in, codes 4415 and 800.4640). These large formats may be of interest for lunar photography, when it is necessary to image the whole Moon, with high general resolution. TP 2415 film is the primary emulsion for astrophotography.

3.6.4 Other black and white and colour films

Ohter films can also be used in special cases. First, Kodak III a J, in 35 mm format, may be of interest for photography in blue and violet light. Kodak Infra-red high speed film may be of great interest for scientific photography. This film is sold in 135 cassette format (36 exposures, code 2481, catalogue No. 1692086). Because of their high sensitivity, the cassettes must be loaded in semidarkness. These films are enormously heat-sensitive and must be stored in a refrigerator below 13 °C (after exposure, they must be stored below 4 °C). The use of such film is very difficult in hot countries: all the infrared films sent to me in Rwanda, by airmail, were fogged (in their original containers), owing to sudden warming by the local conditions. The overall sensitivity of infrared film is hard to measure (without filters it is 80–200° ISO). Its sensitivity extends over the whole of the visible spectrum, from the ultraviolet up to the near-infrared (90 nm). Thus, it is essential to employ filters to eliminate the unwanted wavelengths. For most purposes, a Wratten 25 red filter will suffice to suppress the violet and blue end of the spectrum. Better results are obtained with a Wratten 29 (deep red) filter. Still more selective filters can sometimes be useful: W70, W87, W87C, W88A, W89B. The Kodak IR films have a rather average resolving power: 80 lines/mm for the highest contrast ratio (1000 : 1), and only 32 lines/mm for the lowest contrasts (1.6 : 1). One must therefore use the film with a large focal ratio (F/D = 150–200) in order greatly to amplify the prime focus image and so avoid a loss of resolution and a very grainy negative. The Kodak IR films should be developed in active developers to obtain the contrast desirable in high resolution astrophotography. The most popular seems to be D19, with a development time of at least 6 min, at 20 °C, which gives a γ of about 1.70, indispensable in planetary photography.

For solar work, in certain cases, Agfaortho 25 professional or Kodak Ektagraphic HC may be preferred. Being orthochromatic, they help to eliminate the secondary spectrum of certain refractor objectives if the appropriate filters (W61 or W74) are used. Other films also might be of some use: Kodak Kodalith ortho 2556 type 3 (a slow orthochromatic film, very contrasty) and Kodak fine-grain positive film 1776. These two films could be used for copying or the duplicating of unsharp masks. The first is intended for contrast enhancement; the second has a more general use.

When I found myself in Cotonou (Benin), between 1980 and 1984, I made much use of Ilford XP1 400 film, a very special film, having a high sensitivity and a very fine grain and a high exposure latitude. Based on the principle of colour films, XP1 400 is developed in the same way as colour negative films (the Kodak C41 process) or by using the special Ilford XP1 developer. This film allowed me to make very pleasing paper prints of lunar photographs which offered a wide range of grey scale values. Unfortunately, the contrast of the film remains very low, and its resolving power is poor. It cannot reach very high resolution in lunar work and I do not recommend using it. One must accept using TP 2415, developed to a γ of not less than 1.6, even if the negative thus obtained becomes hard to print on paper.

I am hardly a supporter of colour photography when it comes to high resolution work: colour films (be they colour negatives or transparencies) have inadequate contrast and inferior resolving power. I know that large-circulation astronomical magazines and many of their readers want colour photographs of the Solar System (although the Sun and Moon are essentially monochrome!). One can therefore do as everyone else does and try, for example, Kodak Ektachrome 200 film (with which D. Parker has obtained very good results), and force the development. The new Ektachrome 100, more for the professional, guarantees a higher colour saturation but its sensitivity is wanting. Kodachrome 200 can be tried for the planets, Kodachrome 64 for the Moon and even Kodachrome 25 for photographing the lunar phases with very high definition.

In my opinion, there is only one way in which high resolution colour photographs having high contrast can be obtained: by using black and white negative film (TP 2415) and colour filters. The object is photographed three times successively through suitable blue, green and red filters and the final image obtained by compositing in the darkroom: the so-called 'trichrome' process. According to Wallis & Provin (1988), the most useful filters are the Wratten 47 (blue), W57A (green) and W23A (red), but other combinations could be envisaged. However, trichrome photography and compositing is not a method for beginners! (For more details, see Acker, 1987.)

3.7 Handling the films

Opinions concerning the contrast needed for obtaining the maximum resolving power with TP 2415 are extremely diverse. On maximum-contrast subjects (which are used to estimate the resolution) the optimum resolving power of 2415 is attainable with non-energetic developers. In astronomical photography, however, the finest details which we want to record have low contrast (because they are at the limit of the telescope's resolving power, diffraction effects have the effect of reducing the contrast almost to zero, representing the practical limit for the visibility of detail). Even on the Moon, the smallest craterlets or the finest rilles appear, telescopically, with very low contrast. Energetic developers can therefore be used, in order to increase the contrast of the image projected on the film, even if the resolving power of 2415 is adversely affected (this is not serious: it is sufficient to increase the focal ratio F/D to further enlarge the image). Certain writers have indicated that Technidol LC can be useful for 'pictorial' photography on TP 2415 film. This developer is to be avoided, for the images obtained show that the contrast is very low and the sensitivity of the film falls to only 25° ISO. (One could, however, use it for photographing the whole of the lunar disk, so long as the film is developed for at least 25 min at 20 °C, giving a CI of about 1.0 and a sensitivity of 50° ISO.) Kodak recommend D76 developer for all work other than that needing maximum contrast. By overdeveloping up to 12 min at 20 °C, a contrast (γ) of 2.5 can be reached, with a fine grain and very high resolution. Unfortunately, the sensitivity of the film thus processed will not exceed 135° ISO, which is insufficient for the Moon and planets. D76 developer will be useful for solar photography, when it will benefit from having been exposed to plenty of light. Dobbins *et al.* (1988), recalling the variation of light intensity of the countryside illuminated by the Sun (the brightness ratio going up to 1:500; see also Adams, 1948), regard planetary photography as a special case of everyday photography, and consider that energetic developers must not be used, in order to avoid losing details in the grey scale. The film cannot record a high range of intensities, when it is developed to a maximum CI of 0.8 (γ about 1.2). I disagree, since the range of brightness on Jupiter and Saturn, for example, does not exceed 1:20 or 1:30. If a low contrast development procedure is used, only a greyness is obtained on the film in which the various light levels are confused. The American authors also forget that when we approach the limiting resolving power of a telescope, the contrast of the finest details tends towards zero. We therefore have to differentiate the grey tones and make the smallest details visible, by a development procedure which offers much higher contrast than usual.

Starting from the principle of harmonious, or 'compensating', development, Dobbins *et al.* (1988) recommend the use of Agfa Rodinal developer, at high dilution (up to 1:100) but with considerable overdevelopment. These authors give, on p. 177 of their book, a comparative table of contrast values, sensitivities and resolving powers, obtained with TP 2415 film and various development methods. Some of the data in this table are exact and correspond to the values which can be found in technical documentation by Kodak, but others seem highly improbable. I established sensitivity curves, obtained from the same parameters (types of developer and duration of development) and obtained different figures, notably concerning the contrast obtained with Rodinal developer. I therefore prefer to return to the beginning and to study the working of the main developers used in astrophotography. Therefore, the data that I shall be giving are based either on documentation from Kodak or on my own densitometric measurements (made with a Macbeth 932 digital densitometer). Measurement of the resolving power is quite hazardous, varying very little, between 70 and 110 lines/mm, for low contrast subjects. The determination of sensitivity (ISO) is also very approximate, and must be considered as an order-of-magnitude estimate. On the other hand, the contrast can be measured precisely. I prefer to use the old γ notation, which is very easy to measure, as the characteristic curves for TP 2415 film have a large linear part. I was thus able to check the γ data given by Dobbins *et al.* or by Eastman-Kodak. For the values of the CI, I found that the figures stated by Kodak were the most reliable. For an equal contrast, the values of the CI are lower than the γ values. Having begun the most detailed study of the various developers useful in high resolution astrophotography, I recalled that we seek, above all, to obtain the highest possible sensitivity, a very high contrast (γ ranging from 1.2 to 3, according to subject) and the minimum of chemical fogging. The form of the characteristic curve, notably at the shoulder, plays an important role in the recording of low brightness levels. Those wishing to have the most detailed general background information on this subject should consult the following: Photo-Lab-Index (Morgan & Morgan), Adams (1948), Glafkides (1976), Eggleston (1984), Clerc (1939), Eaton (1965), etc.

3.7.1 The principal developers used in high resolution astrophotography

Kodak HC-110 This universal developer is very interesting, as it can be bought in concentrated solution which can be kept indefinitely. I do not recommend buying so-called 'stock solution' which has already been diluted, as it does not keep so well and is more expensive. This latter solution can be made (as needed) by adding three volumes of water to one volume of the concentrate. I find that there is no difficulty in preparing the various dilutions needed from the concentrate. It will be enough to measure exactly the quantity of concentrate into a small graduated measuring cylinder and, after tipping it into the final graduated container, to wash out the small cylinder half a dozen times with water and add the washings to the final container. These diluted solutions, ready for use, keep badly and must be used quickly. For high quality work it is preferable to use the developer once and to throw it away after use. Some development tanks use no more than 250 cm³ of developer; with a litre of stock solution four 36 exposure films could therefore be processed.

Very dilute solutions are convenient for general photography. In astrophotography, concentrations A and B (see Table 3.10) are to be preferred. In practice, for high resolution astrophotography, either dilution A (63 cm³ of concentrated developer and enough water to make 1000 cm³) or, particularly, dilution B (32 cm³ concentrated developer and enough water to make 1000 cm³) is used. To obtain dilution B (1 : 31), one could also add 7 cm³ of the concentrate to 250 cm³ water, a convenient amount for a development tank of that size, or add 16 cm³ of concentrate to 500 cm³ water for a larger tank. The development capacity is not great: a maximum of four 36 exposure 35 mm films per litre. That is to say, it is preferable to use four separate portions of 250 cm³ (throwing each portion away) than to use 1 l of solution four times. The shelf-life of the HC-110 concentrate is more or less indefinite, but it is best to keep the bottle in darkness, in a cool place. The reserve solution also keeps well. Dilution A can be kept for 6 months in full bottles but only for 2 months if air is present. These shelf-lives fall, respectively, to 3 months and 1 month for dilution B, and the solutions should therefore not be diluted until they are required for use. HC-110 acts rapidly, gives images having a good acutance, an average grain and a very variable contrast according to the dilution and development time

Table 3.10 *Recommended dilutions of HC-110 (after Kodak, with the kind permission of Kodak-Pathé)*

Dilution required:	No. of volumes of water to be added to:[a]		Remarks
	1 volume of concentrated solution	1 volume of stock solution	
A	15	3	Maximum contrast
B	31	7	Optimum concentration
C	19	4	Intermediate concentration
D	39	9	
E	47	11	Of little interest in astrophotography
F	79	19	

[a] All volumes are in cubic centimetres.

Table 3.11 *Formulae for Kodak D19 and D19b developers*

D19		D19b	
Water	500 cm³ (at 50 °C)	Water	500 cm³ (at 50 °C)
Metol	2 g	Metol	2 g
Sodium sulphite (anhydrous)	90 g	Sodium sulphite (anhydrous)	72 g
Hydroquinone	8 g	Hydroquinone	8.8 g
Sodium carbonate monohydrate	52.5 g	Sodium carbonate monohydrate	48 g
Potassium bromide	5 g	Potassium bromide	5 g

After dissolving the solids, add enough water (lukewarm) to make up to 1000 cm³ of solution.

(see Table 3.12, p. 52). It makes very good use of the maximum sensitivity of 2415 film. It is certainly a universal developer, adaptable for the majority of high resolution astrophotographic work. The development times run from 5 to 12 min at 20 °C (see Table 3.12).

Kodak D19 and D19b These developers are of the energetic type, much used in deep-sky astrophotography, since they produce negatives of very high contrast with no fogging. Unfortunately, they make less use of the sensitivity of TP 2415 than does HC-110. D19 can be made by the photographer, according to the recipe (D19b is a variant of this recipe) given in Table 3.11.

These very high contrast developers keep for 6 months in full bottles (with no air), but only for 2 months if even a little air is present. It is advisable to prepare only 1000 cm³ at a time, and to use the solution once only; 250 cm³ will suffice for a single 36 exposure 35 mm film. Development times vary from 4 to 7 min at 20 °C. These developers remain very useful for high resolution photography. Even Dobbins *et al.* (1988) recommend them for Jupiter (5 min at 20 °C), but consider that they do not give much detail in the dark areas. They can also be used for Mars but, in my opinion, they are too energetic for the Moon; the contrast obtained is too harsh (one cannot think of reducing the development time, as that would lead to a loss of sensitivity, a serious drawback). For most photographs at high resolution, in agreement with C. Arsidi and G. Thérin, I recommend HC-110 instead, which is more subtle and makes the best use of the sensitivity of TP 2415 film. But D19 can give real service in solar photography (in hydrogen alpha

light and for details of the solar granulation) and sometimes also for Jupiter, when working with a red filter and a telescope of at least 400 mm aperture. Another surprising thing: a still higher contrast with D19 can be obtained if it is diluted in a 1:2 ratio; by developing for 7 min at 20 °C one can thus obtain a contrast index of 2.70 and a γ of about 3.70 (Kodak data).

When we wish to increase the sensitivity of TP 2415 film and at the same time to have high contrast, we can try Kodak 50-19a, a variant of D19, in which the following are added to a litre of D19:

Kodak Anti-Fog No. 2 (0.2% solution) (to make this solution dissolve 2 g solid in 1000 cm³ water)   20 cm³
hydrazine dihydrochloride (hydrazonium dichloride) (caution! this is a skin irritant)   1.6 g
cold water to make up to a final volume of 30 cm³

30 cm³ of the above antifog solution is added to 1000 cm³ of D19a developer. The mixture must be made up for immediate use. At 20 °C the development time may be extended up to 10 min or more. A significant increase in density will be obtained with a fog level not exceeding 0.40. This may be useful for some types of Jupiter photography in cloudy conditions or at high magnification, yet still with short exposure times.

MWP2 (Difley, 1968) This developer was formulated at Mount Wilson observatory as a replacement for D19. It has the advantage of having less effect upon the sensitivity of the film

Table 3.12 *General summary of sensitivities and contrasts obtained with TP 2415 film according to the method of development (from data by Eastman-Kodak, Dobbins et al., 1988, and original work by the author)*

Developer and dilution	Development time (20 °C)	CI	γ	Relative sensitivity (ISO)
Kodak Dektol	3 min	2.50	3.60	200°
Kodak D19	4 min	2.35	2.90	125°
Kodak D19	8 min	2.50	3.50	200°
Kodak D19 (dil. 1 : 2)	7 min	2.70	3.70	160°
Kodak HC-110 (dil. B)	6 min	1.45	1.70	160°
Kodak HC-110 (dil. B)	8 min	1.70	2.05	200°
Kodak HC-110 (dil. B)	12 min	2.10	2.70	250°[a]
Kodak D76	12 min	2.10	2.60	140°
Agfa Rodinal 1:50	12 min[b]	1.05	1.25	150°
Agfa Rodinal 1:50	14 min[b]	1.10	1.35	200°
Agfa Rodinal 1:25	14 min[b]	1.40	1.70	250°[a]

[a] Denotes the development method guaranteeing the maximum sensitivity.
[b] Agitated once only in the middle of the development time.

when changing the development time (in order to alter the contrast). Its formula is as follows:

Water	750 cm³ (45 °C)
sodium sulphite (anhydrous)	105 g
hydroquinone	10 g
phenidone	0.40 g
benzotriazole	0.60 g
potassium bromide	2 g
sodium carbonate, anhydrous	30 g

The sodium sulphite and hydroquinone should be dissolved before adding the phenidone; when everything has been dissolved, make up to 1 l (1000 cm³) with more water. By developing for 7–12 min a high contrast, high density negative will be obtained. Designed for deep-sky photography, MWP2 could be used instead of D19, but it has been little used in our field of interest, and so we know little of its effectiveness for high resolution work.

Kodak Dektol Dektol is normally used for developing prints. It has been found to be very effective with TP 2415 film also, which it develops very quickly, giving very high contrast. It is readily available in the form of a powder, in quantities to make up 1 l (or more) of solution. It may be replaced by Ektaflo developer, which is easier to use (being sold as the concentrated solution, ready to dilute). It could be interesting to try it when seeking to obtain an infinite γ value with TP 2415 film (solar granulation, chromosphere in hydrogen alpha light).

Kodak D76 This common developer is considered to give fine grain, and allows useful contrasts to

be obtained with 2415 film, when developed for 12 min at 20 °C. Unfortunately, the loss of sensitivity with this developer is very great and it is not advisable to use it except for sunspot photography (for which high enlargements are envisaged). D76 can be bought in any photographic shop and its formula is well known. Less energetic, 1 l of D76 will not develop more than four 36 exposure 35 mm films.

Kodak DK 50 It is possible to substitute D76 for DK 50, which retains a great delicacy of grain, so as to obtain better contrast and to make better use of the sensitivity of the film. One litre is enough to develop five reels of standard 35 mm film.

Agfa Rodinal This developer was introduced by Dr Andersen in 1888. It is a developer using para-aminophenol (4-aminophenol), sodium sulphite and potassium hydroxide. It has the advantage of being sold in concentrated solution form, is inexpensive and has a long shelf-life. The degree of dilution of Rodinal plays a very important part, as it can exert a compensatory action in overexposed areas of film, provided that the film is agitated very little during development. This developer is characterised by its high acutance and by an excellent rendition of details in underexposed areas. It enables a very variable contrast to be obtained, according to the dilution and development conditions chosen. An unopened bottle of concentrate keeps indefinitely. After opening, it can be kept for several months only, so it is best to buy it in small containers. The developer is used in highly diluted form, 1:25, 1:50 and 1:100, and portions of it are used once only. Rodinal is the favourite developer of two of the great planetary photographers: D. Parker and I. Miyazaki. I used it myself at Flagstaff (Lowell Observatory, USA) and at St. Clément (France) but I did not find it superior to HC-110. It is, in my view, worthwhile using for the Moon or for Mars, but I find that the contrast it gives is too low for Jupiter. Again, in my opinion, the contrasts given by Dobbins et al. (1988, p. 177) seem too low. For Mars, Parker uses Rodinal at 1:100 dilution and develops for 14 min at 20 °C, but he hardly agitates the film at all in the development tank; he considers that a sensitivity of 125° ISO is thereby attained with TP 2415 film, which seems low to me. For Jupiter and Saturn, he recommends a dilution of 1:25 with a development time of 14 min at 20 °C (the sensitivity then rising to 300° ISO(?!)). Miyazaki obtains superb photographs of Jupiter by using Rodinal at a dilution of 1:50 (which seems the most appropriate to me) for 12 min at 20 °C. I myself used Rodinal 1:50 at Flagstaff, on Mars and Jupiter; with a development time of 14 min at 20 °C I found that the negatives showed much more chemical fog than with HC-110 (even at dilution A) and that the sensitivity of TP 2415 film was less well utilised. On the Moon, a dilution of 1:50 and a development time of 12 min at 20 °C allows a very considerable contrast to be obtained. Densitometric measurements allowed me to obtain the characteristic curves for TP 2415, developed in Rodinal or in HC-110. According to my measurements, Rodinal offers more contrasty images than Parker found (Table 3.12). Rodinal, although slightly less energetic than HC-110, enables us to obtain good lunar and planetary photographs, and shows shadow details extremely clearly. But this developer makes less use of the sensitivity of TP 2415 compared with HC-110. Its trump card is to prevent the negative becoming too dense in the overexposed parts (in lunar photography, for example), provided again that we take care not to agitate the film at all throughout its development. If the dilution is high (1:50 or more), the developer reduces the level of darkening in the overexposed parts but continues to work on the underexposed parts: a useful compensating effect.

Discussion

The reader will demand to know, with good reason, how experienced astrophotographers are able to use such very different techniques and yet still obtain excellent photographs. In fact the choice of a method is made gradually: it is the fruit of one's personal experience over a period of time, based upon a few particular factors – the type of turbulence encountered, the degree of contrast given by the telescope available, the value of focal ratio F/D chosen, the quality of the drive, the type of enlarger used, etc. The type of print which is to the photographer's taste is also a factor: are prints which have plenty of grey tones (admirable to look at but hard to publish) preferred or those with high (or even violent) contrasts where the greys have become dark or nearly black?

Of all the developers that I have described here, there are but three in current use: D19 (or D19b), HC-110 and Rodinal. With the last two one can experiment with the dilution factors (and thus the development times). It can therefore be understood that the astronomer who is working in bad seeing will be interested in reducing the exposure time to a minimum, by reducing the focal length and choosing a developer which will push the sensitivity of the film to a maximum. If, further, he enlarges his negatives with an enlarger having low contrast (diffused) lighting, and he likes to have very contrasty pictures, he will choose HC-110 and develop them for longer than normal, thus forcing the very high contrast he needs. Another photographer, enjoying good conditions of seeing with a telescope with little central obstruction and an excellent drive, will prefer to use a less energetic developer, and will be free to use a longer exposure. If he owns an enlarger with contrasty (directed) lighting and on the whole prefers lower contrast prints, he will use Rodinal in preference, diluted to 1:50 and developed to give average contrast. In fact, the only thing that counts is the final result – not the technique chosen to get that result!

We are therefore not surprised if, from one astrophotographer to another, the focal ratio varies from 50 to 200, the exposure times vary from 1 to 8 s, and the developers used range from D19 up to Rodinal diluted to 1 : 100! Each must find the technique that suits. For the beginner, I would advise starting off with HC-110, dilution B and a development time of 12 min (at 20 °C). Techniques will change very gradually, in the light of growing experience.

With regard to development technique, it is understood that the developing tank must be completely inverted 2–4 times every 30 s throughout the duration of the development time. It is always essential to agitate in the same way if consistent results are to be obtained. On the other hand, for Agfa Rodinal, if one wants to get the most from its compensating action (described earlier), one must agitate hardly at all, perhaps just once or twice in all. Of course, it is essential to agitate at least once in order to remove air bubbles from the film which would otherwise spoil the results. The Rodinal results of Table 3.12 were obtained using just one agitation sequence of the tank (6–8 inversions, half-way through development). By constant vigorous agitation throughout the entire development time, it is possible to get really high contrasts (CI almost 2.0 and γ almost 2.65). Because Rodinal is a very dilute and non-energetic developer, it will not have the same action upon a series of 36 minute planetary images as upon 36 views of the Moon's surface (the latter being rich in overexposed regions). Thus, a slightly higher contrast will be obtained on the planets than will be indicated by my densitometric measurements, which sample a rather large area of negative. Contrary to the views of Dobbins *et al.* (1988), the contrasts obtained by overdeveloping in Rodinal are very high. That is why this developer is well suited to high resolution work, in spite of the chemical fog it produces. It should not be forgotten that the 'normal' development time, at 1:50 dilution, is only 6 min at 20 °C (for CI = 0.60).

For high resolution photographs, so hard to obtain, one cannot be content with approximations. All the developers listed here must therefore be used once only, then thrown away. (Hence the need to have small development tanks, if possible holding just 250 cm³.) Some authors are of the opinion that distilled water is indispensable in the preparation of developers. This is no problem, but I do not think it is necessary: even if the domestic water supply is contaminated with solid particles, they can easily be filtered out. Many photographers wet the film prior to development, by washing it with water to which 2–3 drops of wetting agent (Teepol) have been added. This can be considered necessary for very rapid developers such as Dektol; otherwise it is superfluous. On the other hand, it is very important to develop at 20 °C. (A standard thermometer is recommended; although digital thermometers read to 0.1 °C, they may be inaccurate by 1.5 °C, and should therefore be distrusted!) Use, if possible, mercury thermometers designed for use in colour development, calibrated to 0.1 °C. In summer the developer must be cooled down with ice; in winter it will need to be warmed up near a radiator.

3.7.2 Stop-baths, fixing and washing films

It is generally advisable to stop the development procedure suddenly by the use of a stop-bath, consisting of a very dilute solution of acetic (ethanoic) acid (1%, according to Dobbins *et al.*, 1988; 4% according to Glafkides, 1976, and the majority of authors). One can also use Ilford IN stop-bath, which has the advantage of being coloured yellow and turning colourless when it is exhausted. (I find that 1000 cm³ of stop-bath can be used many times.)

The stop-bath eliminates any chalk (calcium carbonate) deposit which might have formed on the surface of the film, prevents the swelling of the gelatine and safeguards the acidity of the fixer. The developer is rather alkaline and, unless neutralised by the stop-bath, would tend to reduce the effectiveness of the fixer; the use of a stop-bath therefore allows the fixing solution (which is reusable) to be kept for longer.

The fixing process consists of dissolving up any silver halide crystals from the photographic emulsion which has not been reduced to silver by the developer. It is therefore an important operation, essential to the long-term conservation of the negative. There are a great many different types of fixers, with formulae based on sodium thiosulphate ('hypo') or ammonium thiosulphate. It is simplest to use the rapid, concentrated ones, diluting them with water. Ilford Hypam works very rapidly and keeps well, so long as the diluted solution is prepared only as it is needed. The diluted solution can be used more than once. TP 2415 film, being very fine-grained, must not be fixed for more than 3 min (if the fixer is fresh). The developing tank must not be opened before 2 min has elapsed; if the film has become completely transparent, it is left for about another 30 s and then washed with water. On the other hand, rapid films with coarser grain, such as some spectroscopic films or High-Speed Infrared film, will need longer for fixing (4 min). It is possible to fix eight 35 mm films (36 exposures) with 1 l of Hypam fixer diluted to working strength. With TP 2415, if the length of fixation exceeds 3 min, the working solution must be thrown away and another prepared.

Washing with water is essential to leach any fixer from the gelatine layer. If washing is not sufficiently thorough, any thiosulphate left in the gelatine will decompose over a long period of time, to leave an opaque layer of sulphur in the negative, thus ruining it. TP 2415 can be washed quickly: it will be enough to change the water in the washing tank every 2 min (washing with running water is not recommended, for a number of reasons). The temperatures of the developer, the stop-bath, the fixer and the first portion of washing water must all be equal: 20 °C (± 1 °C only). This is not always easy to achieve. In Africa I made use of a refrigerator in which all the sensitive emulsions could be stored very conveniently, and all the solutions could be kept at 10–12 °C. The refrigerator contained bottles of all the working solutions, as well as a bottle of cold water, so that a single visit to the fridge followed by warming the solutions to 20 °C was all that was needed to begin work. During development (especially with the slow-acting Rodinal), the other solutions would slowly be warming up: stop-bath at 21 °C perhaps, fixer at 21.5 °C, first washing water at 22 °C, but this was not inconvenient, as it was a gradual increase up to the ambient temperature of the tap-water (which might be 28 °C!). The washing of TP 2415 film can therefore be rapid, especially if the water is warm. It is a good idea to add 3–4 drops of Teepol or Kodak Photo-floo in the final film-wash and to wipe the film carefully with a special film 'squeegee'. It only remains to hang up the film and leave it to dry, which can be done quickly in a warm, dry place. It should dry without any limescale deposits (in the form of small white spots) from the tap-water.

3.8 Examination of the developed film: choosing negatives for printing

This is one of the most fundamental operations of high resolution photography. Owing to the atmospheric turbulence, it is necessary to take a very large number of photos (up to 120 exposures per session for lunar work, or 20–36 exposures for a planet), from which, in general, a single one will be the best (exceptionally the yield may reach 3–4%, but very often the entire product of a photographic session is rejected).

The astrophotographer must set aside plenty of time for the examination of each negative under magnification. A negative viewer (or light-box) must be installed in a comfortable position so that the film can be examined with a high quality magnifying lens. Viscardy (1987) uses powerful watchmaker's magnifying glasses which can be worn like a monocle. Parker, who works with very high focal ratios, is content to use magnifiers giving ×3 and ×5. Personally, I have three magnifying glasses: ×4, ×10 and ×22. The ×4

Schneider lens allows the whole 24 × 36 mm frame to be viewed without deformation, but is of no interest for high resolution work. The ×10 aplanatic Peak lens is the main one used for examining TP 2415 negatives. It is perfectly adequate for planetary photography, but recourse to the × 22 Peak lens is necessary to choose between the highest quality lunar and solar negatives. This examination is very important and, once again, it is necessary to have plenty of time for it. The sharpest negatives (meriting enlargement) are marked with a superfine indelible Stabilo-OHP pen. On the negative which follows (or precedes) the best one, one can write the date and time in writing large enough to be seen in the darkroom. Example: 1989 July 12d 22h 10m UT (if the negative is very dark, a tiny label, 12 × 17 mm, can be attached). There is no reason for cluttering up one's files with discarded negatives, and my wastepaper basket receives 90% of all my negatives! Those thought suitable are cut into strips of four or six views (of which, in general, only one is suitable) and slid into negative envelopes. In a notebook, one can keep a record of all the details about the technique used (telescope, F/D, exposure time, development, etc.). The negatives can be stored in a cool, dry, dust-free place.

3.8.1 Making prints of the selected negatives

The choice of enlarger plays an important role, as it determines, in part, the development technique for the exposed film. Theoretically, the ideal enlarger for high resolution must be the type called à éclairage en lumière dirigeé in de Vaucouleurs et al. (1956), or a condenser enlarger with point light source, which we shall refer to as fully directed lighting. This comprises a clear light bulb, of low wattage (30 W = 6 V, 5 A, for example) of the type used in microscopy, with a filament in the shape of a single tight spiral. This point light source should be placed in line with the condensing lens assembly (at least two short-focus condenser lenses are needed). As in the case of what is termed 'Kohler' illumination in photomicrography, it has to form the image of the filament in the plane of the diaphragm of the objective. Under these conditions an extremely bright and contrasty projected image will be obtained. The resolving power of the objective of the enlarger will be used to the full. An enlargement secured with this type of illumination shows the granulation of the film with extraordinary precision, which ensures that the intrinsically high resolving power of TP 2415 film is used. Unfortunately, this type of enlarger is not available commercially, as it has several serious drawbacks: it is necessary to alter the distance between the light source and the condenser as soon as the enlargement ratio is altered; every grain and every minute defect of the negative are spectacularly highlighted; it is impossible to obtain an even illumination if the objective is considerably stopped down; etc. I was able to use just one enlarger (Valoy de Leitz, modified by Tiranty of Paris), which guaranteed an extraordinary resolution, but the brightness was so high that one had to use dense neutral filters. This type of enlarger was designed for making giant enlargements, 2 m across, from 24×36 mm negatives. Moreover, using it was a lengthy and complicated procedure.

Enlargers most generally make use of what is called *semidirected lighting*: an opal lamp, of average size, illuminates a large double-condenser lens. The lighting level is very good and the image has a normal degree of contrast. Less grain is visible and the defects of the negative are less noticeable, but there is a slight loss of resolution (the objective is working under poor optical conditions).

Nowadays, however, we find plenty of enlargers working with *completely diffused lighting*: a low voltage lamp, with a tiny filament, provided with a parabolic reflector to concentrate the light in a sort of light-box, which diffuses the light in all directions. This lighting arrangement, designed to allow the use of dichroic filters of small size (for making colour prints), is very powerful, and has a tendency to reduce the visibility of the granulation and the faults of the negative. On the other hand, the image loses both contrast and resolution. It therefore seems to be the case that the owner of an enlarger of this type will have to be content with much less contrasty prints than one who has an enlarger with semidirected lighting. The difference in terms of resolution is only slightly noticeable. The details recorded on TP 2415 film, using focal ratios between 60 and 150, show a resolution of from 20 to 8 lines/mm only (whereas TP 2415 resolves about 100 lines/mm, so that any loss of resolution, with low contrast subjects, owes much to the enlarger's own objective). That amounts to saying that a maximum resolution of 60 lines/mm can be counted upon, effectively 3–7 times more than exists in reality. On the other hand, an enlarger working with completely diffused lighting definitely limits the maximum resolution of a

negative obtained with a Schmidt camera working at $F/D = 1.5$ or 2.

I chose a Leitz Focomat V 35 Autofocus enlarger, the lighting system of which is apparently diffused but which has two condensing lenses in the latest model. This enlarger does not have quite such powerful illumination but has the advantage of being able to accept an 'Ilford Multigrade module' (issued in collaboration between the establishments of Leitz-Wild and Ilford) which allows one to obtain very easily – thanks to a simple button which inserts two interference filters – on Ilford Multigrade paper all the contrasts between 0 and 5, and all the intermediate values. One can also – and this is very valuable – obtain different contrasts over different parts of the same positive print.

Whatever type of enlarger is chosen, it is important that it be capable of enlarging the image up to at least ×15, as otherwise one would have to turn it round on its column and project the image onto the floor to obtain a larger image. The quality of the objective is of critical importance, and it is not prudent to buy a cheap one. The WA-Focotar 40 mm $F/D = 2.8$ Leitz lens may be the best on the market, but there are plenty of other excellent ones by Rodenstock, Schneider, Nikon, etc. A symmetric six-component enlarging lens would seem adequate to me. According to the type of objective, the highest resolution will be obtained between $F/D = 5.6$ and $F/D = 11$. It can be important to know the exact enlargement used for every print. Astrophotographers use various methods for measuring the exposure time during print-making: by simply counting the seconds (with or without a metronome), or by using various electronic timers such as the Leitz Focometer. In fact, photoelectric cells for measuring the exact exposure time are rarely used, but digital electronic counters, which count down to zero, are very practical.

The correct printing of high resolution negatives is not particularly easy. Besides, generally speaking, most astrophotographers lack the necessary experience to get the most out of their negatives. Above all, one must know how to optimise the exposure time and to make use of all the necessary 'dodges': pieces of black card, attached to a steel rod, intended for covering certain underexposed parts of the negative; cards pierced with circular or elliptical apertures allowing one to give extra exposure to parts of the print (for example, in trying to underexpose the limbs of Jupiter). Printing is quite an art, complex and very arbitrary, which can only be learnt through experience. I would strongly advise the dusting of negatives prior to printing, using a combination of an antistatic brush and a squeeze-bulb filled with air. To guarantee a perfect focus, a magnifier of some sort must be used, on the baseboard of the enlarger. I use an excellent Micromega Critical Focuser but there are many other designs (at various prices) on the market. They are indispensable. When focusing the projected image, always focus onto a sheet of white paper of the same thickness as the printing paper, or an unexposed, fixed sheet of printing paper.

Although some photographers are prejudiced against resin-coated papers, I use the RC Ilford Multigrade III paper*, which is ideally suited to my multigrade module, installed on my Leitz Focomat enlarger. When printing on resin-coated papers, one must remember that when dry, they slightly change their tone: their density increases a little and their contrast decreases. The print, as viewed in the washing tank, must therefore appear a little lighter and a little more contrasty than seems desirable. The new Ilford Multigrade developer will develop resin-coated paper in a little more than 60 s. Fixing does not take more than 3 min (and in preference should be done in two fixing-baths for 1.5 min each); washing in cold running water takes just 4–5 min. Above all, resin-coated papers must not be allowed to soak for too long. They can be dried in air, using an improvised dryer, but all the same it is preferable to use a warm-air dryer, for drying the prints gradually as they leave the washing-tank (without interrupting the printing process). When they first appeared on the market, resin-coated printing papers (with contrast varying according to the filtration system used) such as the Kodak Polycontrast RC and Ilford Multigrade, obviously had serious faults: the 'black' tones were not very black, the grey tones were not rich enough, and the whites were rather dirty. Since then, very great progress has been made. Purists can use either baryte paper coated with Ilford Multigrade emulsion (richer in grey scale tones, according to some) or normal papers (baryte support) in the usual printing grades used in astrophotography: 2, 3, 4 and 5. A final word: contrast grade 5 obtained with Multigrade papers and filters is a little less contrasty than on resin-coated or baryte paper, where the contrast is fixed at grade 5. It is therefore always useful to have a box of grade 5

paper ready for those negatives which need the highest contrast.*

3.8.2 Special techniques for black and white enlarging

Compositing is a very old technique which has long been used in printing planetary photographs where the negatives have been very grainy. It is a question of superimposing 3–12 negatives (of the shortest possible exposure times and of closely comparable quality) onto a positive film (integration printing), by exposing each negative in an appropriate manner (perfect superimposition of each image, and each negative given an exposure inversely proportional to the number of negatives being used). This technique of compositing entails a considerable reduction in the grain, which translates into an increase in resolution and contrast. It was thanks to this method that it was possible to make considerable progress in planetary photography during the wonderful years from 1940 to 1956 (Pic du Midi, Flagstaff). In our times this technique is rarely used, as we have access to films of very fine grain, and also because we are much lazier than our predecessors! However, compositing also involves numerous inconveniences, sometimes serious ones: in order for the technique to work well, at least five or six negatives of the same quality must be superimposed. It is very rare for as many good negatives as this to be available, even if an entire 36 exposure film is used. Viewed in terms of the standards of high resolution attainable today, superimposition of images demands very great precision, which is hard to achieve, and the expenditure of a considerable amount of time for a relatively small increase in quality. My best planetary photographs have always been very few (1–3). It is only when the images are very bad indeed that one can obtain up to seven or eight exposures, of identical quality, but in this case they may not be worth compositing! Wallis & Provin (1988) give some details of the compositing method. A special device to make the enlargements onto must be made, having a heavy wooden board upon which is fixed (with adhesive tape) the sheet of printing paper or (more likely) a large-format positive film. The board has a kind of

---

* The new Multigrade IV is even better.

* Recently I had to change to Ilford Ilfospeed RC (resin-coated) papers (normal fixed grades) from Multigrade, as I found some advantages: there is more light (Multigrade filters are too dark for high contrast work), so that the exposure can be 50% shorter and the image can be better seen for dodging and for other manipulations during exposure. I now use Multigrade papers only for grades 0 to 1 (very uncommon).

opaque cover, which can be folded over the printing paper or left open (like the cover of a book). Upon the cover is fixed a sheet of white paper, onto which the first negative to be composited is projected. The exact contour of the image is then traced out, with a very sharp pencil, on the paper. (The following is a very simple method: in the darkroom, under the safelight, the film or printing paper is taped to the centre of the heavy board and then the cover is folded down over it. The projected image is then centred on the assembly, and the contour of the first negative is traced out on the white paper.) When ready to make the enlargement, the cover is removed and the first negative is projected onto the film or paper: the exposure time t is given by the ratio T/n, where T is the overall exposure time and n is the number of negatives being composited. The cover is then closed again and the tracing of the first image on the paper is used to centre the second image, which is then exposed in the same way as the first. And so on! The procedure can be repeated as many times as there are good images available, provided that they have all been taken within a short period of time: within 5 min for Mars and 2 min for Jupiter. Compositing, diminishing the contrast, may require using a harder printing paper or a film of the 'litho' type. A good composite image therefore represents a great deal of hard work and it will be absurd to make the composite on printing paper. It will be much better to use a positive film from which as many contact prints as are needed can be made later on, quickly and easily. It is probably true to say that the abandoning of compositing by today's astrophotographers is, in some cases, a mistake. In fact, compositing is returning to favour, thanks to the use of CCD receptors, which enable all the image processing and compositing to be done by computer, very quickly and much more simply! The improvement obtained is significant, more especially as it is much easier to obtain a large number of images on video than by conventional photography.

The technique of *unsharp masking* has been used in astrophotography for more than 50 years. The basis of this method is to compensate for the areas of the negative which have become too dark, in order to obtain well-balanced prints and so recover the details which would not otherwise be revealed by the simple technique of 'dodging' during printing. Unsharp masking is much used in deep-sky photography. Recently David Malin has made spectacular use of it (Malin & Murdin, 1984).

In unsharp masking a positive copy of the original negative is made, on a low-contrast emulsion, under very diffused light. Care is taken that the copy of the negative on the new film is enlarged slightly by an amount equal to the thickness of a thin sheet of glass or even by the thickness of a negative, in order that the mask obtained will be quite defocused. The positive copied under such conditions is generally less dense and less contrasty. It will be placed in contact with the original negative (the emulsion of the mask facing the non-emulsion side of the original), so that the two are precisely superimposed. The sandwich thus obtained is then enlarged onto printing paper (focusing, of course, on the negative and not on the unsharp mask!). The technique is delicate, and has been explained in detail by Malin (1977). Although Kodak Litho-Pan-Masking 4733 film automatically gives an unsharp copy, most astronomers use Kodak Professional 4125 Copy film, developed in HC-110, dilution E (see also Wallis & Provin, 1988, and Acker, 1987). I have, in particular, used Eastman-Kodak fine-grain positive film. De Vaucouleurs *et al.* (1956) specify that the diffuse lighting source should have a high surface area (that is, it should be an opal light bulb), and that it is preferable that it be decentred with respect to the 'sandwich' negative/positive, which should be placed upon a turntable (such as that of a record player) in order to allow it to be moved during the exposure (which has to be lengthy). In exactly superimposing the negative and positive, it is useful to place two tiny crosses on two opposite sides of the back of the original negative with fine Letraset transfer lettering. The two tiny crosses appear white on the mask, so that superimposing them in the sandwich will be very easy. It only remains to tape the sandwich together. An infinite number of unsharp masks, of various contrasts, can be obtained. Each will give a different print, and one of them will perhaps give the desired final result. Several attempts will therefore be needed to get it right. Each type of negative requires a particular density and contrast for its ideal unsharp mask.

In high resolution photography the role of unsharp masking remains rather limited, but it can be of service when printing lunar photographs where the negatives are too contrasty (Viscardy, 1987) or for improving the visibility of certain planetary details where the contrast is too high. Tomio Akutsu printed, by unsharp masking, one of my Jupiter photographs taken at Pic du Midi: the result is very interesting (Fig. 4.44:6). I then used the same method to print my best-ever Mars photograph, also taken at the Pic: it will be found that the use of the unsharp mask alters the brightness and modifies the details in the lighter, desert areas (Fig. 4.38). However, it is certainly not a technique to be used all the time, for each good image.

When making paper prints of high resolution negatives, this harmonisation of the image becomes a very important operation which is often found to be difficult. One must get into the habit of properly using the various types of mask or, even more simply, one's hands to shade different parts of the print which would otherwise be overexposed. Mars is the easiest to deal with, while Jupiter needs more care and attention, in order to avoid printing the limb too dark (whether phased or not). For measurement of the positions of the atmospheric features, it must be made obvious where the borders of the disk are, against the blackness of the background sky. The harmonisation of lunar photographs is particularly difficult to attain, and the same can be said for printing solar negatives in hydrogen alpha light. I shall return to this topic in the second part of this book.

The enlargement of the negative must be sufficient for all its details to be easily seen and, if necessary, measured. However, it must not be blown up too much or the grain will be obvious. The usual enlargements run from about ×5 to ×15, with an average around ×10.

3.9 Copying colour positives

I have already stated that I do not actually like high resolution colour photography. When a successful colour slide has been obtained, there is no question of using the unique original for eventual publication. Therefore it is necessary to make coloured copies of the original. This is not very difficult. To guarantee high resolution, one should avoid using the low-cost accessories sold by various photographic shops, designed for the easy 1:1 copying of colour positives. There are two methods for duplicating a colour slide: (1) by direct printing (slide in contact with the copying film); (2) by photographing the slide by photomacrography in a 1:1 ratio. The second method is easier but not as good as the first in terms of resolution. To duplicate a film successfully, a very good photomacrographic objective is needed, such as the Micro-Nikkor

55 mm $F/D = 2.8$ or one of its equivalents by Leitz, Canon, Olympus or Pentax. Again, a good bellows system is needed which will allow of various enlargements around the ratio 1:1. Using the lens in this way reduces the focal ratio by 2 and thus increases the exposure time fourfold. The Micro-Nikkor 55 mm has a maximum resolution at $F/D = 8$. However, as a result of its use in 1:1 copying, its focal ratio becomes $F/D = 16$. Resolution tests show that even at $F/D = 16$ the results are still acceptable, while allowing a better depth of focus. The colour positive is placed in a slide-holder fitted with an opal glass window, and the whole is illuminated by means of a colour enlarger head equipped with interference filters. One or more opalescent plastic sheets further diffuse the light and allow the exposure time to reach an acceptable value of about 10 s. The most frequently used mounting consists of a column supporting the photomacrography bellows, having a reflex camera at its upper end and at its lower end the special lens stopped down to $F/D = 8$ (first opened to $F/D = 2.8$ for focusing, of course). The colour positive is evenly illuminated by the enlarger head placed above it. It must be checked that the distance between the colour positive and the light source is always the same, to within a few millimetres. It is possible to enlarge slightly the image and copy only part of it, leaving off the perimeter, perhaps. Kodak Ektachrome Dupli No. 5071 film, which is specially optimised for tungsten lamps and for exposure times of several seconds, can be used for the copying process.

The most difficult task is to find the combination of filters needed for the best results: it is possible voluntarily to modify the main colour of the original also. Two filters, the anti-UV and the anti-IR, must be in constant use. Some commercial outfits are designed for rapid and convenient duplication, by means of an electronic flash (and Kodak Ektachrome 50-366 film, which is optimised for an exposure time of just 1/1000 s). The Bowens Illumitran is a very practical piece of equipment to use. Often, 2 or 3 films have to be sacrificed before the right filtration for a 30 m spool of Dupli 5071 film is found. The filtration varies a little when different types of colour films are copied (Ektachromes, Fujichromes, Kodachromes, etc.). Of course, duplicate films must themselves be developed by the E6 process. For more details about copying colour positives, Mike and Pat Q (1978) should be consulted.

Part 2

High resolution photography of the main Solar System bodies

After the long general technical introduction, we must now see how the various techniques can be applied to each celestial body in turn.

4.1 The best times to photograph the Sun, Moon and planets

The best season of the year for solar photography is between April and October, since that is when the Sun is highest above the horizon (for northern hemisphere observers) and when the weather conditions (high pressure during summer) are at their best. The best time of day is two hours after sunrise. In winter the Sun is south of the celestial equator and the best images will be obtained about midday, when the weather is stable and a little misty. In the equatorial regions, between the tropics, the Sun can be observed equally well all year round.

The *Moon* needs to be observed under the best conditions. In our latitudes (40–50° N) the best photos are obtained when the Moon has a high northern declination. Again, in the equatorial regions, the Moon can be photographed the whole year round, while in the southern temperate latitudes the best conditions will be at times of high southern declination.

In our part of Europe, and North America, the most favourable times coincide with winter and spring for photography of the first quarter of a lunation, or summer and autumn for the last quarter. The best high resolution lunar photographs are generally taken in August and September (high altitude, good weather and high pressure generating good images). The months of June and December are mostly unfavourable.

The planet *Venus* is best seen at the times of its greatest elongations, between 40% and 60% phase and corresponding apparent diameters of 30″ and 20″.

For high resolution work, the planet *Mars* should be photographed over a period of 2 months either side of opposition, i.e. for 4 months only during each apparition. (However, for the benefit of organisations wishing to receive Mars observations, observers are encouraged to take routine photographs for longer than this.)

In our latitudes, Mars is only favourably placed when its disk diameter is small, since its declination is at best only slightly north at the time of its closest approaches. Above all, Mars should be photographed from the southern hemisphere (1999–2003) or from the equatorial regions.

Table 4.1 *First and last quarters of the Moon between 1995 and 1998 (after Meeus, 1983)*

1995		1996		1997		1998	
First quarter	Last quarter	First quarter	Last quarter	First quarter	Last quarter	First quarter	Last quarter
8 Jan	24 Jan		13 Jan		2 Jan	5 Jan	20 Jan
7 Feb	22 Feb	27 Jan	12 Feb	15 Jan	31 Jan	3 Feb	19 Feb
9 Mar	23 Mar	26 Feb	12 Mar	14 Feb	2 Mar	5 Mar	21 Mar
8 Apr	22 Apr	27 Mar	10 Apr	16 Mar	31 Mar	3 Apr	19 Apr
7 May	21 May	25 Apr	10 May	14 Apr	30 Apr	3 May	19 May
6 Jun	19 Jun	25 May	8 Jun	14 May	29 May	2 Jun	17 Jun
5 Jul	19 Jul	24 Jun	7 Jul	13 Jun	27 Jun	1 Jul	16 Jul
4 Aug	18 Aug	23 Jul	6 Aug	12 Jul	26 Jul	31 Jul	14 Aug
2 Sep	16 Sep	22 Aug	4 Sep	11 Aug	25 Aug	30 Aug	13 Sep
1 Oct	16 Oct	20 Sep	4 Oct	10 Sep	23 Sep	28 Sep	12 Oct
30 Oct	15 Nov	19 Oct	3 Nov	9 Oct	23 Oct	28 Oct	11 Nov
29 Nov	15 Dec	18 Nov	3 Dec	7 Nov	21 Nov	27 Nov	10 Dec
28 Dec		17 Dec		7 Dec	21 Dec	26 Dec	

Table 4.2 *Greatest elongation of Venus, 1995–2005 (after Meeus, 1983)*

Date	Elongation
1995 Jan 13	46° 58′ W
1996 April 1	45° 58′ E
1996 Aug 20	45° 50′ W
1997 Nov 6	47° 08′ E
1998 Mar 27	46° 30′ W
1999 Jun 11	45° 23′ E
2001 Jan 17	47° 06′ E
2001 Jun 8	45° 50′ W
2002 Aug 22	46° 00′ E
2003 Jan 11	46° 58′ W
2004 Mar 29	46° 00′ E
2004 Aug 17	45° 49′ W
2005 Nov 3	47° 06′ E

Table 4.3 *Oppositions of Mars from 1995 to 2005 (after Meeus, 1983)*

Date	Declination	Diameter
1995 Feb 12	+18° 10′	13″.85
1997 Mar 17	+4° 40′	14″.20
1999 Apr 24	−11° 37′	16″.18
2001 Jun 13	−26° 30′	20″.79
2003 Aug 28	−15° 49′	25″.11
2005 Nov 7	+15° 54′	20″.17

Table 4.4 *Oppositions of Jupiter, 1995–2005 (after Meeus, 1983)*

Date	Declination	Diameter
1995 Jun 1	−21° 14′	45″.54
1996 Jul 4	−22° 53′	47″.05
1997 Aug 9	−16° 40′	49″.70
1998 Sep 16	−4° 11′	49″.46
1999 Oct 23	+10° 00′	49″.70
2000 Nov 28	+20° 26′	48″.63
2002 Jan 1	+23° 01′	47″.03
2003 Feb 2	+17° 43′	45″.51
2004 Mar 4	+7° 38′	44″.50
2005 Apr 3	−4° 03′	44″.19

Table 4.5 *Oppositions of Saturn, 1995–2005 (after Meeus, 1983)*

Date	Declination	Diameter
1995 Sep 14	−5° 36′	19″.37
1996 Sep 26	−0° 50′	19″.61
1997 Oct 10	+4° 05′	19″.86
1998 Oct 23	+8° 56′	20″.10
1999 Nov 6	+13° 27′	20″.31
2000 Nov 19	+17° 21′	20″.49
2001 Dec 3	+20° 18′	20″.62
2002 Dec 17	+22° 02′	20″.70
2003 Dec 31	+22° 25′	20″.70
2005 Jan 13	+21° 20′	20″.64

Jupiter can be photographed every year. However, Jupiter does spend periods in southern declination, during its summer oppositions, and so is best observed from our latitudes during winter.

The planet *Saturn* is best observed under our fine autumn and winter skies by reason of its positive declination at such oppositions (1998–2005).

4.2 The Sun

4.2.1 White-light solar photography

The Sun is the most difficult object to photograph, at high resolution, because of the strong diurnal turbulence it causes in the Earth's atmosphere. It is therefore in this high resolution domain that we see the greatest divergence in the quality of results obtained by amateurs compared with those obtained at the great solar observatories.

4.2.2 Diurnal, local and instrumental turbulence

In general, diurnal turbulence is 4–10 times higher than that found at night. Keenan (1952) considered that one would be obliged to observe for several successive months, for several hours a day, to have just 2 days upon which a resolution of 1 arcsecond might be attainable (the resolving power of a 12 cm OG). Bray & Loughead (1964) constructed a completely automatic heliograph, made from a 12.5 cm refractor, which photographed the Sun continuously, for 5–6 h per day. Thanks to a 'seeing monitor' (a photoelectric system that monitored the steadiness of the solar limb), the camera was only operated when the images were calm. The Australian astronomers were able successfully to reach the limiting resolution of their objective only on exceptional occasions: close on 10% of their negatives had this resolution, which is a very good result and better than one might have expected. Most writers agree that there will be steady moments in even mediocre seeing, from which the necessity for a constant surveillance of the Sun in order to achieve the results we want can be appreciated.

Roddier (1978, 1981) studied the problems of turbulence very thoroughly. The complexity of his researches precludes giving the details here. To simplify very considerably, it could be said that Roddier uses the Fried parameter, which Stix (1989) defined as follows: 'The Fried parameter is the aperture of an imaginary diffraction limited telescope used under the conditions of the actual seeing. As the seeing is variable r_0 depends on the observing site and on time, it also depends on the wavelength.' On this last point, the precise Fried parameter says that r_0 is proportional to $\lambda \times 6/5$, which was experimentally verified. Further, r_0 also depends on the altitude of the Sun, being a maximum at the zenith. Thus, the resolution is limited by the telescope if its diameter D is smaller than r_0, and by the turbulence if D is greater than r_0. Under the usual conditions of solar observation, $r_0 = 4$ cm, but at a good site it might be equal to 8 cm. Fried studied the chances of getting a good-quality solar negative, by taking a rapid sequence of exposures every day. The results depend on the ratio D/r_0 when this ratio is equal to or greater than 3.5. Roddier concluded that, for a ratio of 3.5, the probability of obtaining good images is 0.83. The probability falls to 0.11 for a ratio of 5.0, to 0.02 for a ratio of 6.0 and to 10^{-6} for a ratio of $D/r_0 = 10.0$. For $r_0 = 4$ cm and $D = 24$ cm, we can thus hope to get just one good image in 50, on average. With a 40 cm telescope and with good seeing ($r_0 = 8$ cm), one could expect one good image in 10 (but if $r_0 = 4$ cm this probability falls to just 1 in 10^6!). That explains why solar astronomers often tend to reduce the diameter of their telescopes in the hope of obtaining better images.

In other respects, Stix (1989) asserts that, in the Canary Isles, 'most often 20 cm $< r_0 < 30$ cm and occasionally r_0 is as large as 50 cm'. But these results are based upon a study of the turbulence at high altitude and Stix recognises that 'the radiosonde results do not yet include the seeing caused by ground-level turbulence'.

Most writers agree that the best images are obtained 1.5–2.5 h after sunrise (especially in spring and autumn), as the Sun's rays have not yet had time to heat up the telescope. On the other hand, for those who successfully reduce solar heating by shielding the telescope tube from the Sun's rays, the best images will be found around the time of local noon (as with the Moon and planets).

The quality of a site, in terms of turbulence, depends on a number of factors. The ideal remains the top of a very high mountain surrounded by water, in regions of high pressure (the Canary Isles, for example). In a general way, proximity to water seems a good thing, especially if the Sun rises above the water and the wind is from the sea (blowing away the heated air around the telescope). Hillsides, chaotic terrain, enclosed valleys, surroundings of darker rocks, etc., are the worst sites.

To conclude, the diurnal turbulence varies, on average, from 4″ to 8″ but there are also brief calm moments when the turbulence falls to 2″ or 3″ only. Really stable images last only for fractions of a second: hence the need to take plenty of photographs, with short exposure times. At the best-known sites (Pic du Midi, Tenerife, etc.), the turbulence can fall to 0″.5 or even below. Miracles can also occur: our colleague W. Lille very often obtains, in his small village of Stade, solar photographs of incredibly high resolution (see below).

For high resolution solar photography, it is essential to protect the telescope from direct solar heating. At the Mount Teide Observatory on Tenerife, my Spanish friends open the dome of the 40 cm vacuum solar telescope for only about 20 s, just long enough to take several photos, before closing it again. Then they wait for the telescope to cool before taking further images. It is therefore in the interest of the amateur solar observer to keep his telescope in the shade right up to the instant of taking a series of photographs. One must wait a moment for the temperatures to equilibrate but one should not observe for any length of time. It is then best to shield the telescope again (perhaps with a large parasol coated with aluminised plastic?) and to take another series of pictures later.

4.2.3 Instrumental and optical considerations

It is best to use a refractor (being less susceptible to atmospheric turbulence), provided that the aperture is not too large. A small 60 mm refractor has a resolving power of 2″ and allows spectacular sunspot photographs to be obtained, but the drawback of very small instruments is their relative mechanical instability. The standard instrument for solar photography would have an object glass between 100 mm and 130 mm across. Even the great professionals such as Bray and Loughead worked for several years with a 125 mm refractor. I have photographed the Sun for more than 30 years with various refractors and reflectors ranging from 80 mm to 250 mm diameter, and sometimes in favourable sites, but I have never been able to obtain a single photograph which has a resolution better than 1″! I have worked for more than 2 years with a 178 mm apochromatic refractor; I have obtained only slightly higher resolution with this instrument compared with that obtained with my C14 equipped with an off-centre 110 mm filter.

Father Josset was one of the first to secure solar photographs having a resolution better than 1″

(with a 232 mm refractor). Later, in 1979, F. Rouvière obtained the first really exceptional photographs of the Sun with a 20 cm solar telescope (a modified Newtonian) operated from a high altitude site (Fig. 4.5:2). His photos had a resolution of about 0".7. The German Cord-Hinrich Jahn is one of the world's best solar photographers. He uses the 220 mm refractor of the University of Hannover and his best photos of the photosphere show details as small as c. 0".6 (and sometimes better). To my knowledge, only one amateur, the German W. Lille, has succeeded in photographing the solar granulation with the fantastic resolution of 0".25. Incredible but true! I shall be devoting a special section to W. Lille and his techniques (see p. 144).

For most solar astrophotographers, it is probably not only pointless but also disadvantageous to photograph the Sun with a telescope of 200 mm aperture or larger. The ideal, in my view, remains the 150 mm refractor which – once or twice a year – will allow useful photography of the solar granulation. If the secondary spectrum of the refractor used is significant, it is possible to eliminate it by means of a deep-green Wratten 74 filter. When using a telescope with mirrors whose extremity is open, there is a risk of getting very bad images, as a result of the convection currents set up inside the tube. Therefore, it is necessary to close up the tube with an opaque disk (painted white) having an eccentric opening covered with a suitable rejection filter (see below). The diameter of the filter will be determined by the diameter of the main mirror and the size of the central obscuration (7 cm for a 20 cm telescope, 11 cm for a 35 cm one, etc.). One can also close the tube of a reflector with a full-aperture rejection filter, but such a filter is extremely expensive and, if it is more than 25 cm in diameter, diurnal turbulence will severely restrict the useful aperture (to 6 cm only for most of the time; to 10 cm a dozen times a year; or to 15 cm, or more, only in exceptional, absolutely unpredictable cases).

One must do everything possible to avoid the telescope becoming heated before commencing photography. A plastic tarpaulin covering the telescope is the worst solution of all. One can try to cover it with a sheet of very thin aluminised plastic film. A dome is not a good idea either, because part of the interior will become heated and lead to strong convection currents around the slit or opening. Failing all else, the slit must be opened and closed after each sequence of photographs. To my mind, the ideal solution is a

shed on wheels, well ventilated and shielded from the effect of the Sun, as described on p. 7. When the observer wants to study the Sun for a long period of time (6–7 h per day, perhaps), it will be necessary to protect the whole of the telescope (mounting included!) from direct sunlight, by a large, light, white board or hoarding fixed in front of it, casting its shadow over the entire instrument. (Personally I have not been able to use this sort of shield because of frequent strong winds at my observing site.) One can also use a very large umbrella covered with aluminised plastic film and attached to a heavy trolley on wheels; it should be mobile enough to enable the telescope to be quickly uncovered just for short periods. Of course, it is important to be able to mount the telescope upon a grassy lawn; if one only has a concrete terrace, it should be painted with the new IR blocking paint (see p. 7) to reduce the heat absorbed (and therefore radiated) by the ground.

The elimination of excess light and heat is the main task of the amateur; professionals are able to refrigerate the focal plane by means of a circulatory cooling water system (Rösch, 1950; Bray & Loughead, 1964). The best solution which comes to my mind is to make use of one or more reflections from unsilvered mirrors, either by using a Herschel wedge or by constructing a refracto-reflector, in which the light beam is reflected twice, by means of Duran or Zerodur glass. I made two such 'heliographs': one of 80 mm, the other of 106 mm (see Bourge et al., 1979; Fig. 4.1). The first mirror is unsilvered and reflects about 4% of the light; the second must be very slightly aluminised (reflecting 20–30%), in order to get rid of most of the light and to be able to use selective and strongly tinted filters. Herschel wedges, used in conjunction with very narrow-band monochromatic filters, have produced some of the most beautiful solar photographs known (Josset, Jahn, Lille) but they cannot be used with telescopes using mirrors (Fig. 4.2). That is why the simplest solution (and the best, in terms of internal heating) is to use a rejection filter, placed in front of the objective of a refractor or covering the open end of the tube of a reflector. Some are happy to use a simple sheet of Mylar film, which transmits about 1/1000 of the incident sunlight. This is the ideal ratio for high resolution photography, but for direct viewing of the image on the focusing screen it is necessary further to protect the eye with the aid of an appropriate neutral grey filter (Wratten 96). I have always been sceptical about the optical quality of Mylar

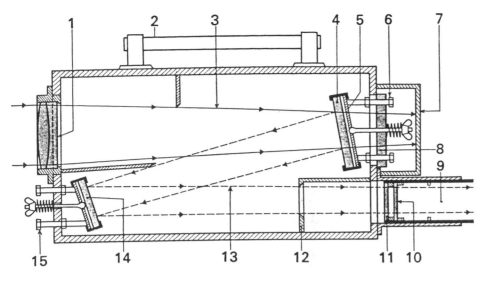

Figure 4.1. Cross-section through a Schaer-type solar refractor, *constructed by the author.*
1. 110 mm objective. 4. Flat unsilvered mirror made of Duran glass. 6. Adjusting screw for the mirror. 8. Plexiglas window to let the sunlight through.

7. Removable cover. 14. Secondary flat mirror, partially aluminised (R = 30%). *15. Adjusting screws. 12. Diaphragm. 10, 11. Colour filters. After Bourge* et al. *(1979).*

Figure 4.2. The great solar refractor of W. Lille (Stade), *comprising a simple 'Chromat' object-glass (C) of 300 mm aperture: 1, general view of this impressive heliograph; 2, the long open tube holding the objective.*

filters, but the solar photographs of G. Thérin have convinced me of their effectiveness, at least for an effective aperture of 10 cm. Otherwise, the ideal remains the aluminised optical glass plane-parallel window. For direct (and completely safe) visual observation, filters whose transmissions range from 1/10 000 to 1/100 000 are to be recommended, but they are not convenient for high-magnification photography. The best transmission for photography is between 1/500 and 1/2000. Reputable firms offer these filters in sizes from 100 mm to 200 mm. I successfully used several filters supplied by Lichtenknecker. Carl Zeiss Jena also supply them. Certain very cheap rejection filters made in the USA should be distrusted, since a window of plane-parallel optical glass, if perfectly made, is more difficult (and consequently much more expensive!) to make than a parabolic mirror of the same diameter. The firms which supply Catadioptric telescopes (Questar, Celestron, Meade, etc.) also offer solar rejection filters consisting of an aluminised glass plate, but their transmission is generally too low for high resolution photography (safety considerations restrict the fabrication of numbers of solar filters which would be considered dangerous for direct visual observation). For further details about these rejection filters, see p. 41.

4.2.4 Photographic considerations

The photography of the whole disk of the Sun can be useful in certain fields. On the normal 24×36 mm format the problem is to obtain a solar image between 19 mm and 21 mm diameter. With a telescope of focal length between 80 cm and 1 m.60, it is possible to achieve the necessary enlargement of the image with the help of a negative doublet or Barlow lens or a teleconverter lens. With a Fluorite Vixen 102 mm refractor, a TC $\times 2$ is most suitable. For my Astro-Physics 178 mm refractor (1 m.60 focal length), I use the TC $\times 1.4$ (apochromatic) teleconverter lens, which gives an image of the Sun about 20 mm across. It is always necessary to avoid using an eyepiece giving a very low magnification. Telescopes whose focal lengths vary from 1 m.80 to 2 m.20 can directly give an image of the desired size. If the instrument has a focal length greater than 2 m.40, a large-format negative must be used (TP 2415 film in 120 roll film format or sheet films 10×12.5 cm–4×5 in).

For photographing the details of individual sunspots, the focal ratio should be increased to $F/$

$D = 50$–100 (100–180 mm refractors), which can be done either by using two coupled $\times 2$ Barlow lenses or by using a high quality eyepiece such as a Plössl, Eudioscopic or projection eyepiece (12–20 mm focus). A microscope objective is also to be recommended (its magnification varying from $\times 4$ to $\times 10$, depending on the initial focal ratio).

Of course, the camera should be a reflex camera, with an interchangeable focusing screen and adjustable focusing. (That is, it should be possible to adjust the viewing lens to focus precisely on the focusing screen.) For lightweight refractors, the Olympus OM1, OM2, OM3 or OM4-Ti seem to be most suitable. They will be equipped with a reticulated transparent glass focusing screen and a Varimagni viewer. Although they are no longer manufactured, one can sometimes come across the OM1, OM2 and OM3. Working with a very heavy refractor, I prefer to use Nikon F2 and F3 cameras, provided with transparent focusing screens and their famous DW3 or DW4 direct viewers, $\times 6$ magnification and adjustable focusing. When working with an OM2 or OM3, OM4-Ti or Nikon F3, it is possible to use automatic exposure in order to obtain negatives with a constant density. For this it will suffice to sacrifice a cartridge of 2415 film, to calibrate the exposure meter, and take several shots for various sensitivity ratings (ISO 100–250° approximately). With my Nikon F3, I obtain the optimum density for a sensitivity rating of ISO 200° (for TP 2415 film, developed in HC-110). According to the altitude of the Sun, the transparency of the atmosphere and the density of the filter employed (I use the Wratten 18*), the exposure times vary between 1/250 and 1/1000 s. One should be especially careful not to use longer exposure times – 1/125 or 1/60 s, for example.

Photography of the Sun requires high contrast films, a high resolving power and ultra-fine grain. Kodak TP 2415 film seems to be to be very convenient, but other films such as Agfaortho 25 or Kodak Ektagraphic can also be used.

The solar photographer must wait for the best moments of seeing. At my site, such moments are found 2 h after sunrise or about local noon. The image being constantly turbulent, focusing is not easy to achieve, and thus it must be constantly checked. One must not be discouraged if the image on the focusing screen seems to be very poor; the photo obtained will often be much

---

* The W18 filter gives good results on sunspots with TP 2415 film, but I prefer a deep green filter and orthochromatic (Agfaortho 25) film for photographing the solar granulation.

better, thanks to the vagaries of the atmospheric turbulence and its rapid changes. It is therefore essential to watch the quivering image closely in the viewer, and not to take photographs until the steadiest moments. An auto-wind motor makes the task easier, as one can concentrate on watching the image. It is necessary to be patient, wait for the brief moments and to expose at least a whole 36 exposure film in the hope of getting at least one useful image. If the group of spots under scrutiny is a particularly attractive one, and if the images seem to be better than usual, one might sacrifice two or even three rolls of film in the hope of obtaining one high resolution image.

The films should be developed in an energetic developer to make the best use of their sensitivity and in order to get high contrast. Three Kodak developers may be used: D19, Dektol and HC-110 (dilution B). Dektol ensures maximum contrast (useful for showing the solar granulation), whereas HC-110 (dilution B) developed for 10–12 min at 20 °C will give well-balanced images. The MWP2 developer (Difley, 1968) or Tetenal's Dokumol can also be used for high contrasts. Photographs of the entire solar disk require less vigorous development: HC-110 (dilution B) for only 6–8 min at 20 °C will be perfect.

4.2.5 Results obtained

The choice of the best negative is a difficult operation, but an essential one. At least two good-quality magnifying glasses and a light-box or negative viewer are essential. The × 10 magnifier quickly allows one to identify the best negatives of the film. A ×16 or ×22 magnifier seems to me to be indispensable for the final selection, as there is generally only one 'best' negative.

A good negative (one in 20, 36 or even 72 exposures) will next be printed onto paper, with an average enlargement of ×10. I use Ilford Multigrade III paper and my interferential module (described earlier) is adjusted for the grades from 3.5 to 5. It is necessary to dodge the large sunspot groups for about 50% of the exposure time by means of a small cardboard disk supported by a steel rod, without which the umbrae and penumbrae will be 'burnt out'. Thus, the exposure is judged for the background disk (photospheric granulation and faculae details). The length of this dodging procedure will vary from one negative to the next. It is also possible to make prints using unsharp-masking (see p. 58), as F. Rouvière has done (Figs. 4.3 and 4.5).

In general views of the solar surface, showing sunspots, the highest resolution obtained is of the order of 4″. Photographs of the details of sunspots give an average resolution from 2″ to 2″.5. At this resolution we can begin to see the striated appearance of the penumbrae and the discontinuous nature of the photosphere. Very rarely (three or four times a year), it is possible to hope for a resolution of 1″.5–1″.2, which allows the structure of the penumbrae and the rice grains (which do not stand out well in ordinary photographs) to be very clearly perceived. However, the image is never totally crisp over the whole negative.

Some amateurs, particularly patient ones and a few lucky ones, have successfully taken solar images whose resolution has fallen below 1″ (Father Josset, F. Rouvière, C-H. Jahn and, above all, W. Lille). (See Figs. 4.4–4.7.)

I am no advocate of colour solar photography. The Sun is yellow-orange (therefore of a single colour only), no more spectacular in colour than in black and white. Colour films have a very low resolving power and a disappointing contrast. Kodak Photomicrography colour film can be tried, or a low-sensitivity Ektachrome, of high contrast and high resolving power, but it is only found in the USA and its development necessitates the E4 process, which very few processing laboratories offer. Otherwise, one can always try the trichrome process, used with TP 2415 film and three complementary filters (red, green and blue). One can also photograph the Sun in black and white (TP 2415 again) and print the negative on colour paper, inserting a yellow-orange filter in the light path!

Figure 4.3. White-light solar photographs. *1. Sunspots and photospheric granulation, 1991 July 28 at 10.15 UT, by G. Thérin, 100 mm OG; Mylar rejection filter; green W58 filter; 1/1000 s on TP 2415 film. 2. Group of sunspots with faculae, 1989 August 31, by D. Lachaud, 100 mm OG. 3. Large sunspot near the limb, 1978 July 20, by F. Rouvière, with a 205 mm Newtonian; glass rejection filter (T = 1/500); F/D = 68; green filter; 1/500 s on Recordak film, developed in D19. This is an unsharp-masked print, to a scale where the Sun's diameter is 165 cm.*

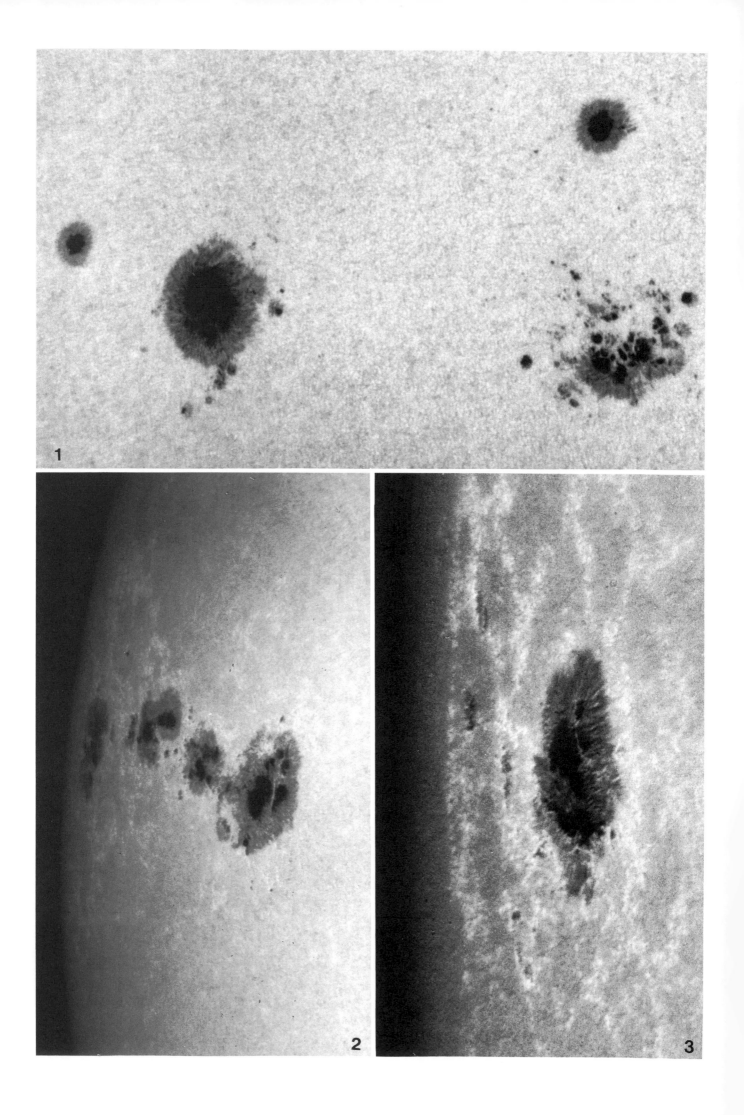

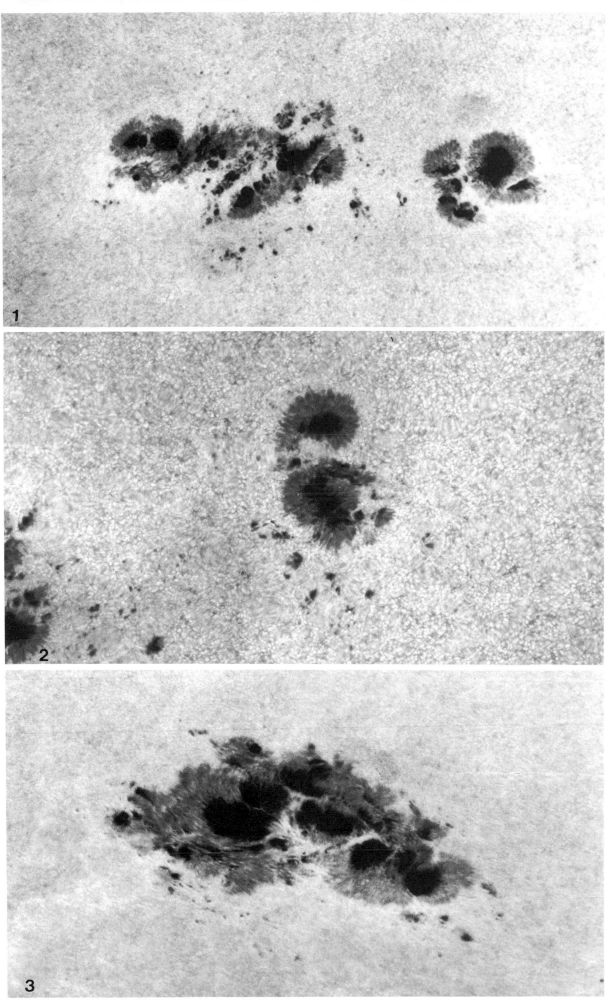

Figure 4.4. Great sunspot groups. *1. 1989 June 11, by the author with a C14, equipped with an off-axis 110 mm rejection filter, on TP 2415 film; 1/100 s. 2. 1988 August 1 at 13.57 UT, by Cord-Hinrich Jahn, with the 200 mm OG of the University of Hannover; Herschel wedge; green filter; 1/500 s on TP 2415 film. 3. 1991 June 12 at 07.50 UT, by the author, with a 178 mm OG; glass rejection filter (1/1000); W21 orange filter; 1/700 s on TP 2415 film.*

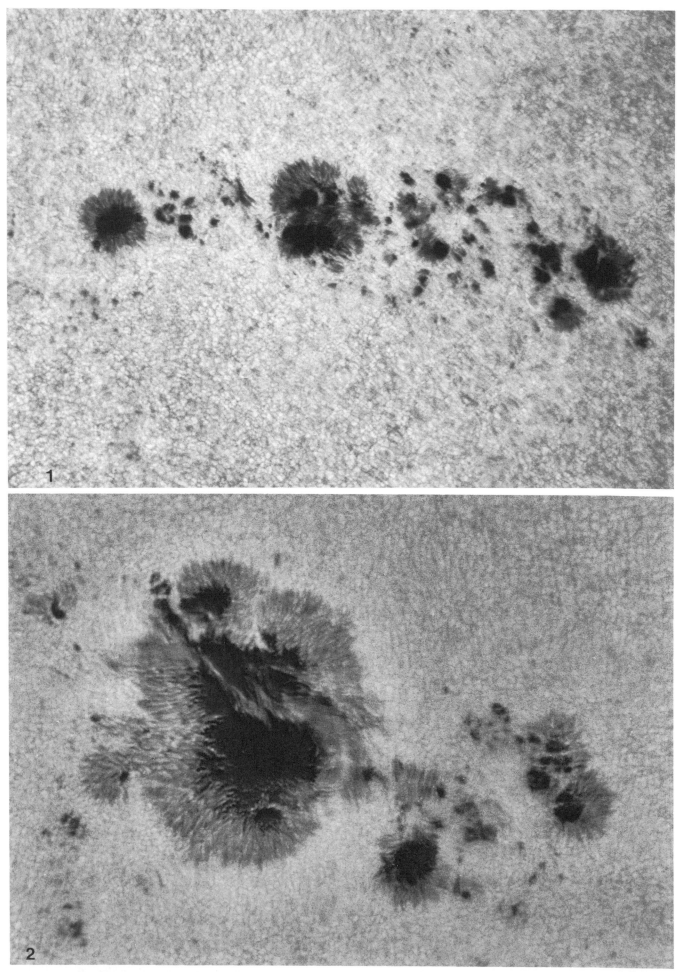

Figure 4.5. The first real high resolution amateur solar work. *1. 1988 May 26 at 14.23 UT, by Cord-Hinrich Jahn, with a 200 mm OG (focal length 13 m). 2. 1978 July 14 at 10.32 UT, by F.* *Rouvière with a 205 mm Newtonian; rejection filter; F/D = 68; 1/500 s; Recordak film; unsharp-masked print. Resolution reaching 0".7.*

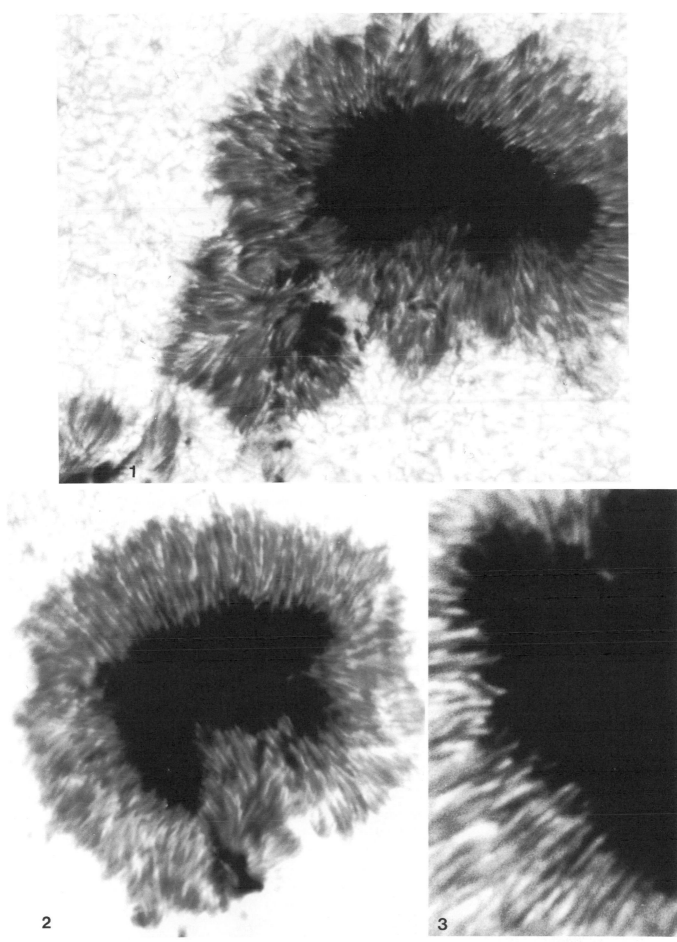

Figure 4.6. High resolution achieved! 1. 1992 May 12 at 16.15 UT, by Wolfgang Lille (Stade), with his 300 mm OG; Herschel wedge; green monochromatic interference filter; 45 m focal length (!); TP 2415 film. 2. 1992 May 21, with the same telescope, but in less good (?) seeing. 3. The same day, with a 45 m focal length. These three images are quite exceptional, their resolution reaching 0".40 or better. Incredible but true!

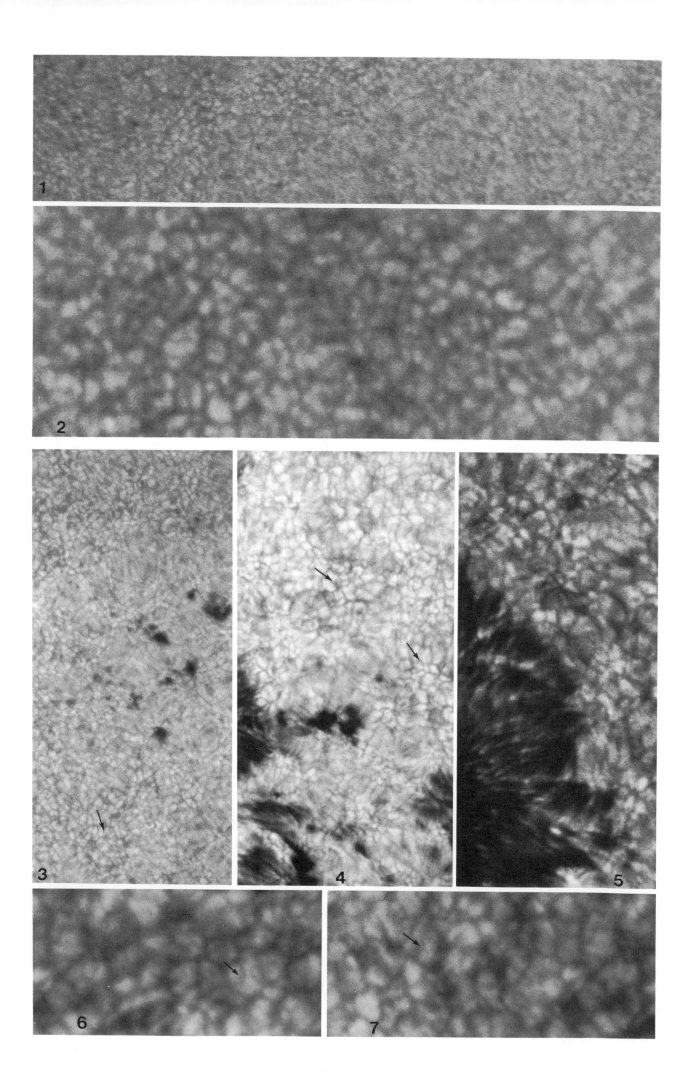

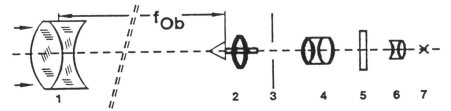

Figure 4.8. The principle of the Baader coronograph. *1. Object-glass, 80–90 mm diameter. 2. Aspheric rejection lens, supporting at its centre the conical reflector which covers the image of the Sun. 3. The iris diaphragm which eliminates the light* *scattered by the edge of the object-glass. 4. Projection lenses, enlarging the image of the occulting cone. 5. Hα interference filter (centred on λ = 656.3 nm, 40 nm bandwidth). 6. Eyepiece. 7. Eye or camera.*

4.2.6 Solar photography in Hα light

Although most amateurs photograph the Sun in white light, some devotees attempt to record it on film in the longer wavelength of hydrogen, especially in Hα (6563 Å = 656.3 nm), which shows chromospheric phenomena: spicules, filaments, active zones, eruptions (flares) and prominences.

Prominences and coronographs
Prominences arouse considerable interest and are generally photographed by means of a *coronograph*. It is possible to construct a coronograph oneself, although it is not very easy (Mazereau & Bourge, 1985). Ready-made coronographs are also available, such as those sold by Lichtenknecker. I purchased a 90 mm coronograph (with a simple, aspheric objective) which works very well, but its interference filter has rather too wide a passband. A Baader filter,

with only a 4 Å (0.4 nm) passband, much improves the image quality and prevents the eye being dazzled during the preliminary adjustments.

Today it is possible to transform any fluorite Vixen refractor, 80 mm in diameter and of 910 mm focal length, into an efficient coronograph thanks to the 'Protuberanzenansatz Mod. II', made by Baader in Munich. This accessory consists of a compact optical assembly, containing six reflecting cones which cover the Sun's image, the aspheric transfer lens, an iris diaphragm, a projection lens and an interference filter with a bandwidth from 10 Å to 4 Å (see Paech, 1989; Fig. 4.8). The module weighs less than 400 g. In fact, the 80 mm refractor gives a solar image that is a little too small (less than 8 mm diameter), so that one cannot expect to reach high resolution, even with perfect images. This formula is ideal for observing and photographing the whole of the Sun, surrounded by its prominences, even if the latter are very large (which is not usually the case), but the specialist needs to be able to study smaller areas at high resolution. Happily, Baader will produce coronograph modules suitable for longer focal lengths (1300, 1600 or 2250 mm) on demand. Some German amateurs have obtained very good results with a Baader coronograph specially designed for the Astro-Physics Starfire 178 mm refractor. For diameters above 80 mm, Baader also make Schott RG 610 rejection filters. These commercial coronographs give a bright image: when photographing the focal plane image, exposures of 1/125 and 1/250 s will be adequate, on TP 2415 film.

For prominence photography, coronographs have the advantage of an affordable price and of giving bright images allowing short exposure times. The Sun is masked by the occulting cone, an effect recalling a total eclipse of the Sun.

Figure 4.7. Solar granulation photographed by amateurs. *1. 1967 March 8 at 14.00 UT, by G. Nemec, with a 200 mm refractor and a helioscope. 2. [no date given] by W. Lille with a simple 'Chromat' objective of 180 mm aperture, Herschel wedge and monochromatic filter. 3. 1988 August 1, by Cord-Hinrich Jahn, with a 200 mm OG (same data as for Fig. 4.5:1). 4. 1992 May 22, by W. Lille with the 300 mm 'Chromat' refractor; 45 m focal length. 5. The same day, at 18.15 UT. 6, 7. 1991 August 3, both by W. Lille, with the 300 mm objective and a focal length of 125 m; 1/30 s on TP 2415 film. Diameter of the Sun on this scale = 9 m! These photographs of the solar granulation are quite extraordinary, their resolution attaining 0″.30 or even better.*

Normally one would use a 24 × 36 reflex camera, with a transparent focusing screen, and TP 2415 film, developing in HC-110 (dilution B) for 6–8 min at 20 °C (for prominences, too high a contrast is a nuisance). It is also easy to photograph the prominences in colour, preferably with a high resolution film such as Kodachrome 25 or Ektachrome 50, or the new Fujicolor G100° films. If the exposure time exceeds 1/250 s, it is necessary to use neutral filters to bring the exposure time up to 1 s, to avoid the vibrations due to the shutter release, so troublesome between 1/10 s and 1/30 s: for details of the full-aperture shutter technique (or 'hat-trick'), see p. 39.

Nevertheless, I prefer to photograph the prominences with the aid of a Hα Day Star filter, which is much more costly than a coronograph attachment but which also allows me to study the chromosphere.

Hα filters
The only firm making high quality Hα filters at affordable prices is Day Star in the USA. For several years, Day Star has offered a new filter, called a 'scanner', the price of which is very reasonable. This filter was specially designed for visual observation, and Del Woods, the director of Day Star, hardly recommends it for photography. The firm also has filters of the 'University' type: they are high performance but very costly. The filters are not guaranteed for more than 5 years, as a result of the inevitable ageing of such multilayer interference filters. In practice, however, they are usable for a good 12 years or so. The notice which accompanies the filter informs us that it is a 'sophisticated instrument'. The filter is not bulky and weighs less than 600 g. It transmits hydrogen alpha light, with a bandwidth of 0.5 Å, 0.6 Å or 0.7 Å (centred on 6562 Å–656.2 nm), depending on the model chosen. The University 0.7 Å filter provides a better view of the prominences but shows less contrast in the chromosphere. The 0.5 Å filter, the most costly, improves the contrast of the chromospheric details, but the prominences now appear rather dull. The 0.6 Å filter therefore remains the most universal and it was the one which I chose to buy. My filter operates at a temperature of 43 °C (the thermostat works at 110 V, for a power consumption of 15 W). The potentiometer that adjusts the temperature is calibrated from 0.0 to 10.0. In the one which I use, one must choose the 8.0 setting to be 'tuned in'. Setting the reading at 7.0, one obtains a wavelength shifted 1/3 Å into the blue, while a

setting of 9.0 shifts the wavelength 1/3 Å into the red. I obtain the maximum contrast with the 8.0 setting, for all the chromospheric features. It is necessary to place a deep-red rejection filter in front of the telescope objective. This filter reflects 80% of the visible wavelengths but transmits 92% of the Hα light.

It would be interesting to know how the Hα filters are made. Unfortunately, the secret has been well kept (taking the unit to pieces invalidates the 5 year guarantee). One can therefore only guess. After having consulted my friend Professor Rouvière and going through various documents, I assume (without certainty) that the Day Star University filter is, basically, a filter derived from the Fabry–Perot interferometer. It involves one or two sheets of glass, whose faces are plane-parallel to an accuracy of $\lambda/20$ or better, and partially metallically coated. As a result of multiple reflections between these two (or four) faces, the solar radiation undergoes quite a series of interference phenomena, and the filter thus transmits several 'crests' of different wavelengths. Multilayer interference filters, then, eliminate the unwanted wavelengths, finally isolating and transmitting just the hydrogen alpha wavelength. Complex filters of this type (like those of the Lyot type, which are still more complex) are very sensitive to temperature variations, so their physical construction should be absolutely rigid. Thus, we are obliged to warm the filter to a temperature above ambient, and hold it at that temperature, as it would be much more difficult and costly to refrigerate it at a lower temperature. My Day Star University 0.6 Å filter works at exactly 43 °C; as I have already said, small variations in temperature can shift the wavelength a little further into the red, or towards the blue. Such filters are relatively fragile, and are therefore best kept in a very dry cupboard or drawer away from light and dust.

The Hα filter, like other monochromatic filters, requires to be illuminated by almost parallel light in order to produce a satisfactory image. In the absence of a built-in optical system of the coupled variety (a combination of two converging lenses) which automatically supplies parallel light, one is obliged to use an objective lens with a focal ratio F/D greater than 30. It is not advisable to use a simple Barlow achromat to obtain a focal ratio F/D from 30 to 40, for the results will be catastrophic (the rays emerging from a negative doublet are divergent). It would therefore be simplest to have an optician polish a simple objective lens of long focal length ($F/D = 35$), which will not need to be aspheric. If we take the diameter of the objective to be 10 cm (the largest in keeping with the limitations of diurnal turbulence), the focal length will be 3.50–4 m, giving a very large solar image, allowing an acceptable resolving power. The mounting of such a lens is not easy: it can be placed either horizontally (using a heliostat) or in refracto-reflector mode (the rectangular tube will now be only 1.20–1.30 m long; see Fig. 4.1). One can even shape the lens directly from Schott RG 610 red glass, which is used to make rejection filters, needed for the protection of Day Star filters.

A good solution consists in using a Cassegrain or Schmidt–Cassegrain telescope of long focal length. I placed my 116-mm-diameter rejection filter on my C14 (355 mm Celestron), in an offset manner to avoid the central obstruction: with a C8 or C11 one must accept having to use a smaller rejection filter, 80 mm or 60 mm, respectively. In this manner, I now have an unobstructed telescope 116 mm in diameter and of 3 m.90 focal length, at $F/D = 34$, which gives a solar image 37 mm across. Under these conditions, the image given by the C14 is quite perfect, like that of a refractor. The rejection filter being decentred, the beam of light impinging on the Day Star filter will be found to be very slightly tilted. It is therefore essential to tilt the hydrogen alpha filter at an equivalent angle. The manufacturer has already provided a bevel-edged ring for this operation (specially calculated for use with the C14; the angle of adjustment would not be the same with another telescope). In these conditions, the rejection filter must be rotated to the 12 o'clock position in front of the C14 (Fig. 4.9). The University filter needs to be oriented accordingly, the widest part of the bevel-edged device directed upward. The use of the Day Star filter on the C14 is very easy and very convenient, as the telescope is extremely compact and has fine adjustments in RA and declination. The 116 mm diameter is enough for high resolution and the long focal length helps, although in the daytime one can hardly hope to do better than attain a resolution of 1".5. I could imagine having a 20 cm $F/D = 35$ Cassegrain made (7 m focus), but

Figure 4.9. Equipment used by the author for Hα solar photography. *1. C14 equipped with its rejection filter (Rf). 2. Day Star University 0.6 Å Hα filter mounted on the Visoflex III reflex unit (V) and the Leica Mda camera body (L) (with a very gentle shutter release).*

I do not think that much would be gained in terms of resolution, at least from my observing site. For 3 years I have scrutinised the Sun in Hα almost daily; I have obtained more than 4000 images, but I think I have never reached the theoretical resolution of my instrument. Usually, the turbulence is so high that the instrument cannot be used at all!

Being retired and able to do what I like with my time, I direct my C14 with its Day Star filter towards the Sun from when it is first visible, and visual and photographic observations carry on from 10 a.m. till 5 p.m. I take photographs of any great prominences, flares or active sunspot groups, whenever they are visible.

Taking photographs in Hα light

For photography with a Hα filter, I use two Leica cameras (chosen for their smooth shutter action), mounted on a Visoflex III 'endoscopic' body (with interchangeable focusing screens), which permits manual operation of the reflex mirror. The vibrations produced by the flipping of the reflex mirror of Nikon cameras lead to a serious loss of resolution with shutter speeds of $1/15$–$1/30$ s, these being the speeds most often used in photographing the chromosphere. The image observed on the clear glass focusing screen is relatively dim, so focusing will therefore be very difficult. Del Woods advises taking several photographs on each occasion, with a few either side of what seems to be the best position of focus. This advice is sensible for beginners. In fact, one can clearly see whether the image is sharp or not; it suffices to refocus after each exposure, first on the graticule of the focusing screen and then on the chromosphere. One needs to take a good 6–9 images to have a chance of getting a usable one (obtained during a moment of steadier seeing). At Saint Clément, the best images are obtained either at about 10 a.m. or between 1 p.m. and 3 p.m., when the C14 is orientated at right angles to the valley. (I find myself situated, unfortunately, on a hillside.) The rejection filter is fixed on a silvered cap, itself protected by another reflective cover. However, the telescope in full sunlight will be certain to overheat. A very large reflective hoarding is needed, fixed in front of the telescope, casting its shadow over the whole instrument. As I have said elsewhere, my site is subject to constant sudden strong gusts of wind, so unfortunately I cannot use such a sunshield. The conditions of thermal equilibrium of my set-up are thus rather bad, continual surveillance of the Sun being disadvantageous to obtaining really good images. (I am thinking of setting up a mobile heatproof shed which can be rolled away each time exposures are taken.)

I use two cameras for taking photographs, one for prominences and one for details of the chromosphere; the films are not developed in the same manner (see later). Exposure times are $1/60$, $1/50$ and $1/30$ s for chromospheric details (on TP 2415). For prominences, the exposure time varies between $1/2$ s and 2 s, for which one uses the full-aperture shutter method; see p. 39. (See Figs. 4.10–4.13.)

Even by choosing interesting subjects and the most favourable moments of seeing, the failure rate is high: after taking a 36 exposure film, I never have more than 2–4 suitable images. The problem is always the same: the loss of resolution due to turbulence.

The film reserved for prominence photography must be developed to an average contrast level: 6–7 min in HC-110 (dilution B). In other respects, the film reserved for taking photos of the chromosphere must be developed to a high degree of contrast. For convenience, I always use HC-110 but at concentration A, for 8 min at 20 °C. Del Woods advises using D19 but I think that Dektol would be better still. As always, the negatives should be examined by magnifying glasses ($\times$10, $\times$20), and only the best ones marked for enlarging later.

Printing Hα photos is quite difficult. For the prominences, I enlarge the negative exactly 15 times. I centre the prominence on a sheet of paper (13×18 cm), taped onto the enlarger's baseboard. I then switch off the enlarger and substitute a sheet of printing paper, which is also taped to the baseboard. I then make the print, trying to give the best exposure for each part of the prominence, by using masks of various sorts and at the same time slightly altering the contrast by means of the Multigrade module attached to the enlarger. At the last moment, I cover the prominence with an opaque mask (cut to the radius of the Sun) and I give a second exposure, after having removed the film from the enlarger, in order to blacken the Sun's disk as it appears during eclipse or through the coronograph (Figs. 4.14:1 and 4.15:3). Printing negatives of the chromosphere is frequently exasperating. Most of the negatives taken through the Hα filter exhibit slightly irregular density and, as they have to be printed on grade 4 or 5 paper, the result can be catastrophic. The print should therefore be harmonised as much as possible by masking the lightest zones of the negative during the printing process. One sometimes has to waste ten sheets of photographic paper before getting a really suitable result. In really difficult instances, the photographer must be satisfied with enlarging a small portion of the negative only.

Colour photography can be attempted, but has several notable drawbacks, since one can hardly obtain the same contrast and resolution that TP 2415 offers. For the prominences, one can try fast colour films (Ektachrome 100 and 200), which can give good results with exposure times of 2–4 s (using the full-aperture shutter method). But one always has the problem of overexposing the rest of the Sun's disk. For the chromosphere, the only solution that I can recommend is to use Kodak Photomicrography colour film, which can be obtained in the USA, but do not forget that this film must be developed by the E4 process. The contrast and resolving power of this film are favourable for recording the chromospheric details but, in view of its low sensitivity, an exposure of about 0.5 s is needed, by means of a full-aperture shutter. An exposure time of this order is somewhat inconvenient, being neither long nor short: it is worthwhile inserting a W96 neutral filter to increase the exposure time up to a few seconds, which will be easier to judge. Even simpler, one can take a normal black and white negative (TP 2415) and then print the negatives on colour printing paper by introducing a coloured filter (dense orange for printing a slide or blue-green (cyan) for printing a negative). There is a whole field of research to be explored here!

Some results

In Hα photography one cannot aspire to the heights of very high resolution. The majority of amateur prominence and chromosphere photographs (either through coronographs or with Hα filters) hardly exceed a resolution of 1".5. Bernd Flach-Wilken obtains very fine prominence photographs, thanks to a coronograph mounted on his 178 mm refractor. Better still, W. Lille has been able to obtain equally good results on the prominences and, above all, on the chromospheric features on several occasions, his photographs having a resolution better than or equal to 0".6, thanks to a non-achromatic single-lens objective (a 175 mm 'Chromat'). For the past year he has been working with his new 300 mm 'Chromat'

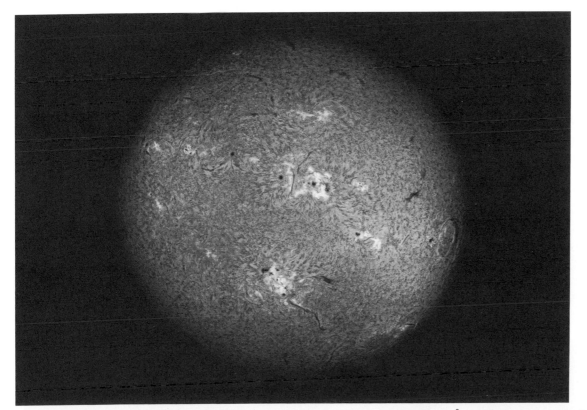

Figure 4.10. The Sun in Hα light *(chromosphere, active regions, filaments, etc.), photographed on 1990 February 25 at 11.03 UT by Franz Kufer, with a 125 mm refractor, diaphragmed to 75 mm and a Day Star filter (0.7 Å bandwidth), on TP 2415 film, 1/60 s. It is not a high resolution image, rather an overall view of Hα phenomena.*

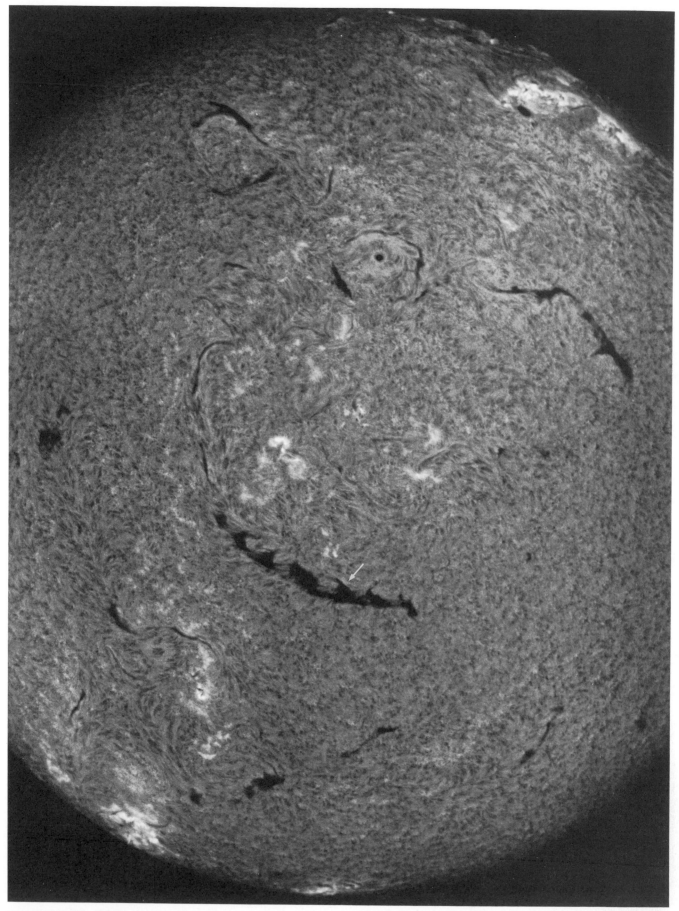

Figure 4.11. Filaments in the solar chromosphere. *1992 November 20 at 10.32 UT, by F. Rouvière with a 80 mm refractor; F/D = 30; Day Star University Hα filter (bandwidth 0.5 Å); 1/30 s; TP 2415 film. The resolution is limited by the diameter of the OG for this wavelength of light.*

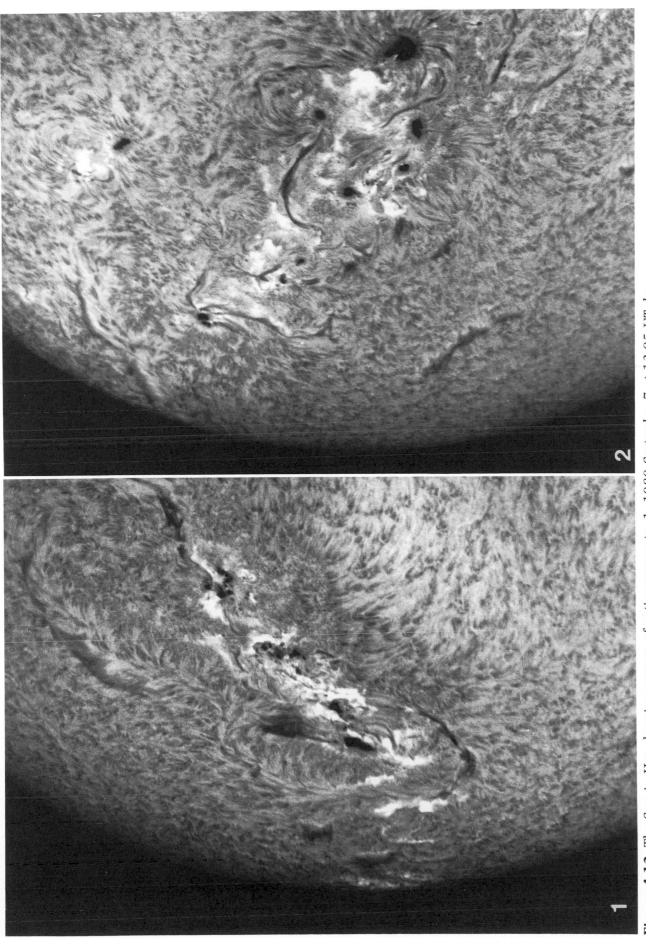

Figure 4.12. The Sun in Hα, showing groups of active sunspots. 1. 1989 September 7 at 13.05 UT, by the author with the C14 and an off-axis 116 mm rejection filter; F/D = 34; Hα Day Star University 0.6 Å filter; 1/30 s; TP 2415 film. 2. 1990 August 23 at 13.35 UT, by the author (same technical data). Resolution limited by turbulence or the wavelength of Hα light.

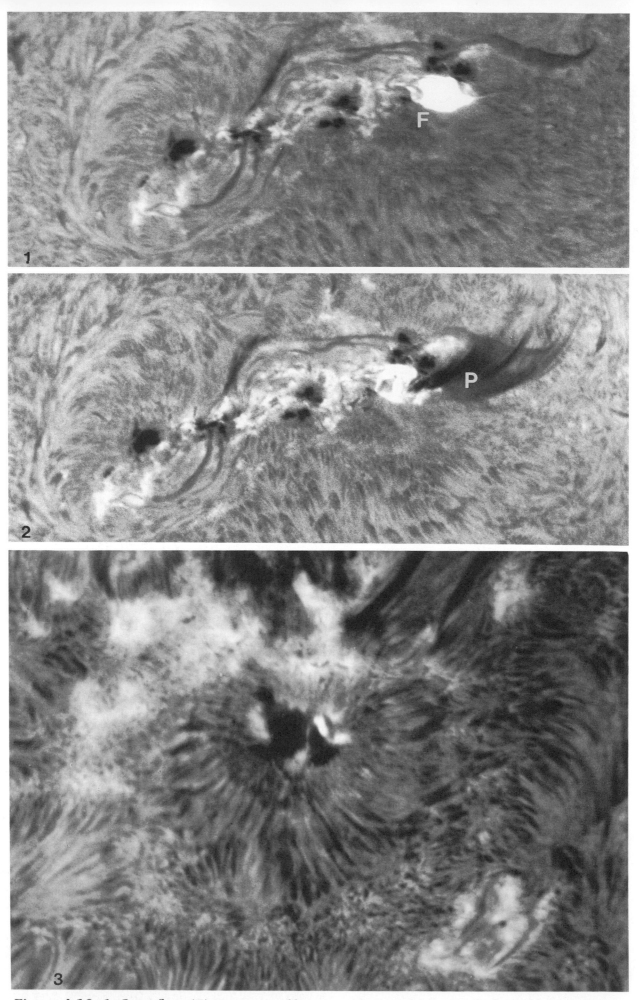

Figure 4.13. 1. Great flare (F) in a group of large sunspots, 1989 September 4 at 09.00 UT, by the author (same technical data as for Fig. 4.12:2). 2. The same, 20 min later (09.20 UT), showing the diminution of the flare and the emission of prominences (P). The resolution of these two photographs is severely limited by atmospheric turbulence. 3. Truly high chromospheric resolution, by W. Lille, with his 180 mm 'Chromat' refractor and a Hα Day Star filter. The resolution reaches 0″.9, which is quite remarkable for this wavelength of light.

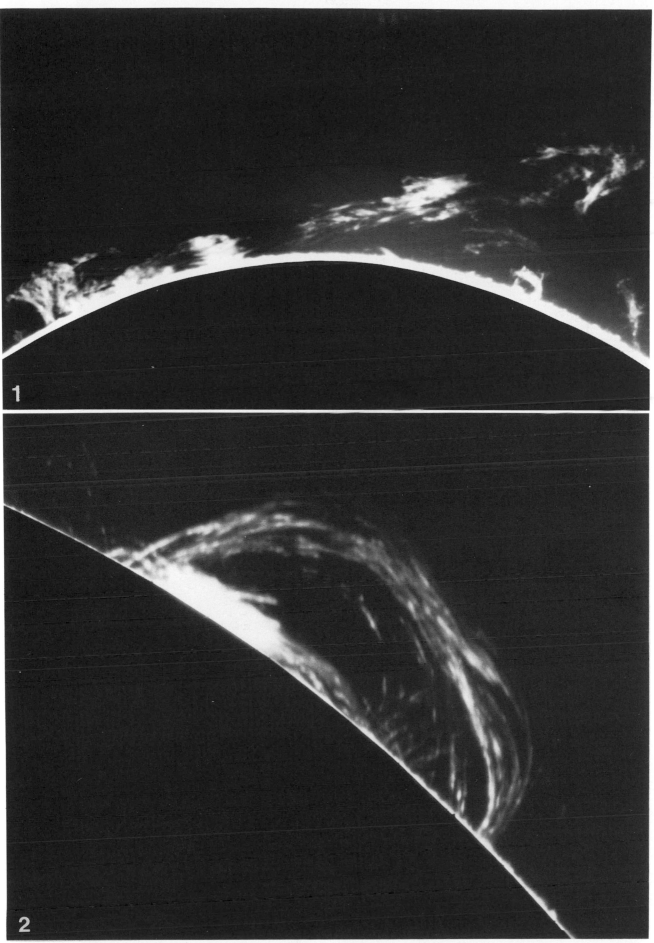

Figure 4.14. *1. Great eruptive prominence, 1990 July 30 at 13.35 UT, by the author with the C14, 116 mm rejection filter and Day Star University Hα 0.6 Å filter; F/D = 34; 1 s; TP 2415 film. A masked print to blacken the disk of the Sun. 2. Superb prominence, 1990 April 1 at 09.55 UT, by Bernd Flach-Wilken (Wirges) with a 178 mm refractor and a Baader coronograph attachment (Hα 3 Å filter). The resolution of this image is exceptional.*

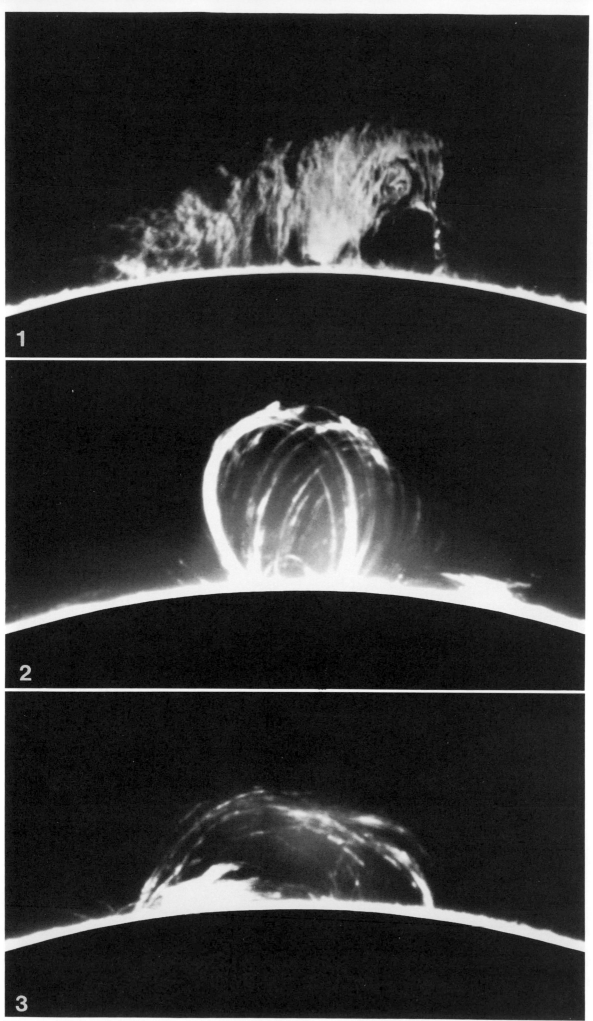

Figure 4.15. Solar prominences, *photographed by the author. 1. 1990 February 12 at 16.10 UT. 2. 1989 August 15 at 10.10 UT. 3. 1990 April 1 at* 10.00 UT. Same technical data as for Fig. 4.14:1. The resolution is average, around 2″, limited by turbulence.

lens, and therefore hopes to obtain even higher resolution.

4.3 The Moon

The Moon is one of the objects most frequently observed by amateur astronomers. It attracts many photographers, who record its unchanging landscape. It is very easy to obtain 'pleasant' lunar photographs, but nearly all are deceptive in their character, because the Moon shows few really fine details. In practice, high resolution lunar photography is very difficult but it is also a unique field where some amateurs have been able to equal or even improve upon the results obtained at professional observatories.

In the early days of astrophotography (1890–1920), the efficiency of photographic techniques was lamentably small; the lunar photographs taken with the 2 m.50 telescope on Mount Wilson showed details which could easily be seen visually with a typical amateur instrument of 25 cm aperture. The best lunar photographs obtained at great observatories between 1930 and 1960 showed an average resolution of 0″.40, which was very poor, such details being accessible with an amateur 20 cm reflector. It must not be forgotten that one should be able to photograph finer lunar details than the theoretical resolving power defined for double-star work (see p. 14; Dragesco, 1969, 1992).

However, the road towards high resolution was a long and difficult one, and the initial results were deceptive (see the lunar photographs in Texereau and de Vaucouleurs, 1954). The first good results were obtained about 1958 by the German amateur Günther Nemec (Roth, 1960), who had built a 20 cm refractor mounted in the manner of Schaer. Nemec's photographs caused a sensation at the time, and many amateurs doubted their authenticity. Their resolution was no better than 0″.6–0″.4, but at the time this was considered extraordinary (Fig. 29). Pope and Osypowski (1969) and H. Zeh in the USA, and then H. Dall and H. R. Hatfield in England occasionally published very remarkable lunar photographs. Since then, progress has been rapid, but only a limited number of amateurs have been really successful, as truly high resolution is very difficult to attain.

4.3.1 Instrumental and atmospheric considerations

One can obtain very bad photos under very good seeing (through inexperience, major technical

errors, etc.), but never vice versa. No technique can be a substitute for perfect images. That is why there are relatively very few high resolution lunar photographers. I shall not attempt to recap here on all the problems (instrumental or local) which can add to the atmospheric turbulence, as these have been discussed on pp. 8, 9. The great advantage of lunar work resides in the unchanging nature of the Moon's surface. One can therefore hope to experience, by chance, several favourable evenings or mornings, perhaps three or four times a year. However, in order to take advantage of such privileged moments, the observer must be ready and waiting at the eyepiece of his telescope: hundreds of hours will be needed to obtain several remarkable photographs. It is an activity for obstinate, indefatigable, even fanatical, amateurs! These can sometimes be found, in favourable locations, as at the 'scoperta' above Monte Carlo where our colleague G. Viscardy (an international pioneer of high resolution) lives, but he too has constantly had to break his own records, good seeing becoming more and more infrequent, particularly with a 52 cm telescope. The publication of Viscardy's monumental lunar atlas was a major achievement, all the more so as in our latitudes, in addition to the problems of atmospheric turbulence caused by the Moon's frequent low altitude, we are often subjected to bad weather, violent winds and excesses of heat or cold. To my knowledge, as I have already stated (p. 45), only certain intertropical regions of the Earth offer conditions which are favourable over the greater part of the year. In Cotonou (Benin, West Africa), the astronomer C. Boyer observed the Moon rising, above the sea, without a trace of atmospheric turbulence, with a 20 cm telescope. The Gabon also seems to be very favourable for good seeing, as are most of the humid countries of the Equatorial regions.

Contrary to what might be imagined, a very large telescope is by no means a necessity for photographing the Moon; we are dealing with a very brilliant celestial object which shows relatively contrasty surface details. Really good lunar photos taken through a fluorite Vixen 102 mm refractor have been published by Arsidi and Ickhanian. A 200 mm telescope seems to be a good size to start with. It should have only a small central obstruction (0.15–0.20) and an excellent primary mirror, better than tenth-wave, if possible. Theoretically, Schmidt–Cassegrains should not be used: they are too complex optically (and so perfection is almost never attained), have too large a central obstruction, and sit upon flimsy mountings. Nevertheless, in spite of a bad press, it was with a C8 (20 cm Schmidt–Cassegrain), then with a 225 mm Schmidt–Cassegrain that G. Thérin was able to take what were to be the best lunar photographs ever obtained with such apertures (with a resolution reaching 0″.25; Figs. 4.17, 4.20, 4.22, etc.). The Schmidt–Cassegrains have several advantages too: they have compact optical assemblies, closed tubes and no spider mounting for the secondary mirror. A diameter of 250 mm should be ideal for the Moon, but the experience of C. Arsidi and B. Flach-Wilken with 30 cm telescopes shows that with good seeing some further gain in resolution is possible. The 400 mm size is restricted to the most exceptional sites (Canary Isles, Florida, Okinawa Island, Equatorial Africa, etc.).

The mounting plays a more important role than might be thought. For lunar photography, polar alignment by Bigourdan's method (or similar) should be carried out as exactly as possible (Martinez notes that, for an alignment error of only 1°, one obtains a drift of 1″ for every 4 s of time). The drive should be as smooth and as accurate as possible. It is easy to monitor the accuracy of the tracking by observing a small craterlet with a power of ×300 or thereabouts, through an eyepiece having a ruled grid or graticule in the field of view (for more details see p. 28). If a steady displacement in RA is seen, it is because the telescope is not being driven at the lunar rate.

If the polar alignment is correct, the drift of the Moon can reach 0″.5 for every second of time, so that an exposure time of 0.75 s should not be exceeded if it is wished to attain a resolution of 0″.35! I therefore find it lamentable that several Japanese telescope manufacturers have omitted to provide for a 'lunar correction' on the handsets which control the stepping motors of their modern equatorial mountings. If what is described as the 'solar' drive rate is used, the drift is reduced and one can have an exposure of up to approximately 1.2 s. With a 220 V synchronous motor, the problem is much simpler: it will be enough to alter the speed of the motor by means of a variable-frequency unit (quartz-controlled), while observing through the eyepiece reticule at high power. During my stay in Cotonou (1980–4), the drive of my telescope was truly unpredictable and I had to change the frequency every five or six exposures. If one has to use an exposure longer than 2 s, one really must also have the means of making a correction for drift in declination. (The Moon's

drift in declination can reach 0″.26 in each second of time.) For various reasons it is always best, as Viscardy (1987) has advised, to keep the exposure to a minimum (1 s or less). One can also obtain high resolution, in bad seeing, with a very long exposure time: 20–60 s! Since, as Canell (quoted by Kopal and Garder, 1974) has found, with the 1 m.50 astrometric telescope of the Flagstaff US Naval Observatory, the undulations of the image due to the turbulent seeing tend to cancel out about a mean position. Unfortunately, this sort of technique requires fantastic precision in the RA and declination slow-motion drives, which is beyond the reach of amateurs.

4.3.2 Optical and photographic considerations

Opinions are divided as to what focal ratios are needed in high resolution lunar work: Viscardy, Arsidi, Thérin and the author use $F/D = 40$–80, and others use $F/D > 100$. A short exposure being the main requirement for success, I concede that a focal ratio of 60 would be high enough for 20–30 cm telescopes (12–18 m focal length) using TP 2415 film.

The enlargement of the prime focus image can be achieved by using two coupled or spaced Clavé Barlow lenses (×2), or Plössl eyepieces or special projection lenses, of 12 mm or 20 mm focal length. The camera should be of the reflex type, 24×36, with an interchangeable transparent focusing screen and an adjustable viewing lens (see pp. 33, 34). With amateur-sized telescopes, exposure can be made using the 'hat-trick' method (see p. 39). It is good to check the focusing and the centring every seven or eight shots. In every way, the slightest movement of the details seen against the eyepiece graticule reduces the chance of obtaining a sharp image, which explains why we hardly ever realise the latter ideal!

Of course, only one emulsion will do for high resolution lunar photography: TP 2415. In the past (Dragesco, 1984) I have recommended XP1 400 film, which exhibits high sensitivity coupled with very fine grain, but unfortunately its contrast and resolving power are insufficient. Very diverse advice has been given concerning the development of TP 2415 film in lunar work. Starting from the idea that the Moon exhibits very contrasty details, Dobbins *et al.* (1988), Martinez (1987) and others have advised obtaining a medium contrast (CI 0.5–0.8), by means of a compensating developer. Technidol, recommended by some, must be rejected out of hand: even when overdeveloped, films will not exceed a CI of 0.7, or a sensitivity of 40° ISO! D76, recommended by Martinez, allows a CI of 1.2 to be reached but the sensitivity cannot be pushed beyond 125° ISO. For the Moon, we must combine maximum sensitivity (in order to keep within the 1.5 s exposure limit) with good contrast. In lunar photography it is too often forgotten that when working at the limit of the resolving power of a telescope, the contrast of the finest details tends towards zero. Our pioneer colleague G. Viscardy used D19, the basic developer of all astrophotography. He pushed the development time up to 7 min, thus getting a very strong contrast ($\gamma = 3.4$; CI = 2.4). However, he finally advocated a development time of 4 min, which somewhat reduced the contrast ($\gamma = 2.90$; CI = 1.90), but this also entailed a loss of sensitivity (the latter being limited to 125° ISO). Negatives developed in D19 are very hard to print. If this was the experience of such a skilled photographer as Viscardy, the task would be almost impossible for a beginner! The best solution would therefore seem to be that chosen by C. Arsidi and G. Thérin: to develop in the universal developer HC-110, dilution B (1/31), for 12 min at 20 °C. Under these conditions, the highest sensitivity possible with TP 2415 (250° ISO) will be obtained, together with a strong but manageable contrast ($\gamma = 2.50–2.70$; CI about 2.10). Printing the negatives thus obtained is always a problem. That is why I also experimented with Rodinal developer. If it is used at 1 : 50 dilution for 14 min at 20 °C, it should be possible to obtain a CI of 1.2 for a sensitivity of 150° ISO. In fact, from the sensitometric graphs which I was able to establish, all depends on the degree of agitation (Rodinal being a non-energetic developer when very dilute). D. Parker recommends using hardly any agitation throughout development. My graphs show that a γ of 1.25 and CI of 1.20 can effectively be obtained if the film is only agitated three times during the development time of 14 min, and a γ of up to 2.0 will result if normal development methods (one agitation per minute) are used. Rodinal having the property of compensating for under- or overexposure, one can obtain a moderate darkening of the overexposed regions and a normal density in the regions next to the terminator, with plenty of detail in the shadows (provided that agitation is strictly limited to once only, after 7 min in the tank). I therefore advise beginners to use either HC-110 (dilution B) for 9 min at 20 °C or Rodinal (1:50) for 12 min at 20 °C with very little agitation, although they will

obtain neither the maximum sensitivity nor the very highest resolution. Of course, all these developers are to be used once only (and thrown away after use).

I must again emphasise the importance of taking plenty of photographs, a 36 exposure film as the absolute minimum per session (and if possible, two or three such films). The choice of the best negative should be made with the help of high-power magnifying glasses ($\times 10$, then $\times 16$ or $\times 22$). The print should be made at an enlargement of $\times 10$ or $\times 12$, preferably on resin-coated paper or Multigrade paper (so that the grade of paper can be varied for the same print, according to the degree of harmonisation required). Finally, producing a really good lunar photographic print demands plenty of work, leads to a significant wastage of printing paper and needs a lot of practice.

4.3.3   Results obtained

I must make a confession, distressing though it is: from 1980 to 1984, when I found myself at Cotonou, under ideal conditions of observation, I let pass the chance to obtain high resolution lunar photographs of exceptional quality, partly for instrumental reasons and partly because I wanted to use Ilford XP1 400 film. Nevertheless, on some negatives taken with TP 2415 and developed (too briefly) in Ilford Microphen (with inadequate contrast), the resolution obtained varies between 0″.60 and 0″.45. (Once only was I able to achieve 0″.30, on the rille in the Alpine Valley: it was an excellent result but hardly an adequate success rate.)

G. Viscardy, thanks to his incredible efforts in France, was able to obtain thousands of lunar negatives with a resolution usually reaching 0″.30, but sometimes 0″.25 or even 0″.15. With his 520 mm reflector, he was thus able to equal or to surpass most of the best lunar photographs obtained at the world's most famous observatories. His *Lunar Atlas* is a masterly work. However, the 520 mm telescope is perhaps too large for its site at Saint-Martin-de-Peille, which seems to have progressively deteriorated (Viscardy, 1987).

It is amazing to record the constant, steady progress that has been made over the course of the last 5 or 6 years by two persevering lunar photographers. C. Arsidi, with a 305 mm Cassegrain, photographed, from the Paris region, lunar details as small as 0″.20. G. Thérin, with the aid of a Celestron 8 (on a very rigid Japanese mounting), then with a Takahashi SCT 225,

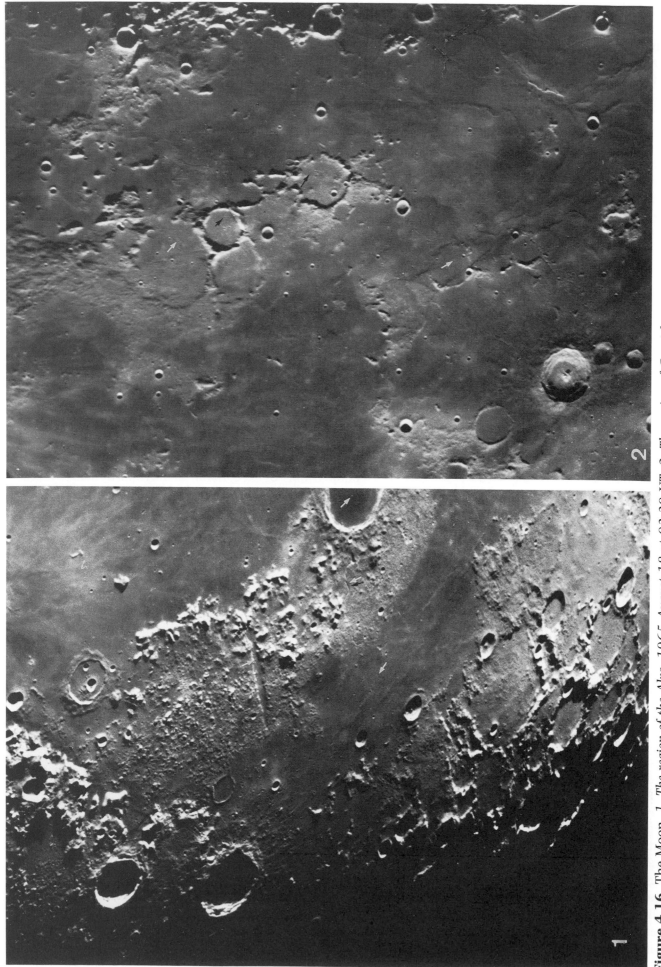

Figure 4.16. *The Moon. 1. The region of the Alps, 1965 August 18 at 03.30 UT. 2. The region of Guericke, Parry, Fra-Mauro, 1965 August 20 at 02.00 UT. Two images by G. Nemec, obtained with his 200 mm refractor; F/D = 70; 2–3 s; Adox KB17 film. These photographs caused a sensation in 1965. The enlargement here is insufficient to show the true resolution of the images (see the arrows), which is, besides, limited by the*

Figure 4.17. The Moon: the crater Fracastorius. 1. 1990 April 1 at 21.00 UT, by C. Ichkanian with a 102 mm apo-fluorite refractor at F/D = 88, on TP 2415 film. 2. 1965 October 14 at 01.10 UT, by G. Neřmec, with his 200 mm refractor (14 m focal length). 3. 1981 September 16 at 01.20 UT, by the author at Cotonou (Benin) with his C14; F/D = 60; 1 s; XP4 400 film. 4. 1992 September 16, by G. Thérin, with a 225 mm Takahashi Schmidt–Cassegrain (F/D = 12; F/D final = 65); 1.2 s; TP 2415 film. Images 2 and 4 show the progress that has been made in high resolution astrophotography over 27 years. (Also compare these images with those of Fig. 4.29.)

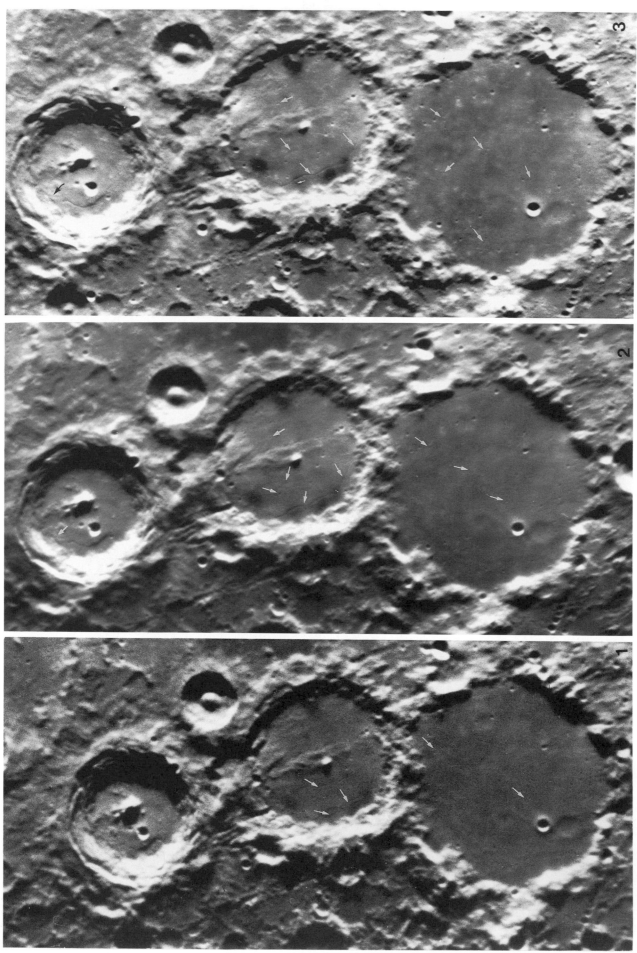

Figure 4.18. The Moon: the craters Ptolemaeus, Alphonsus and Arzachel. 1. 1965 September 8 at 02.30 UT, by G. Neměc with his 200 mm OG. 2. 1988 October 2 at 12.40 UT by the author with the great refractor of the Lowell Observatory (600 mm OG, stopped down to 450 mm): F/D = 45; 0.5 s on TP 2415 film. 3. No date, by C. Arsidi, with a 305 mm Cassegrain, on TP 2415 film. The resolution of this last image is excellent (0".30), and 90 craterlets can be counted on the floor of Ptolemaeus: the small arrows show these details, at the limit of resolution.

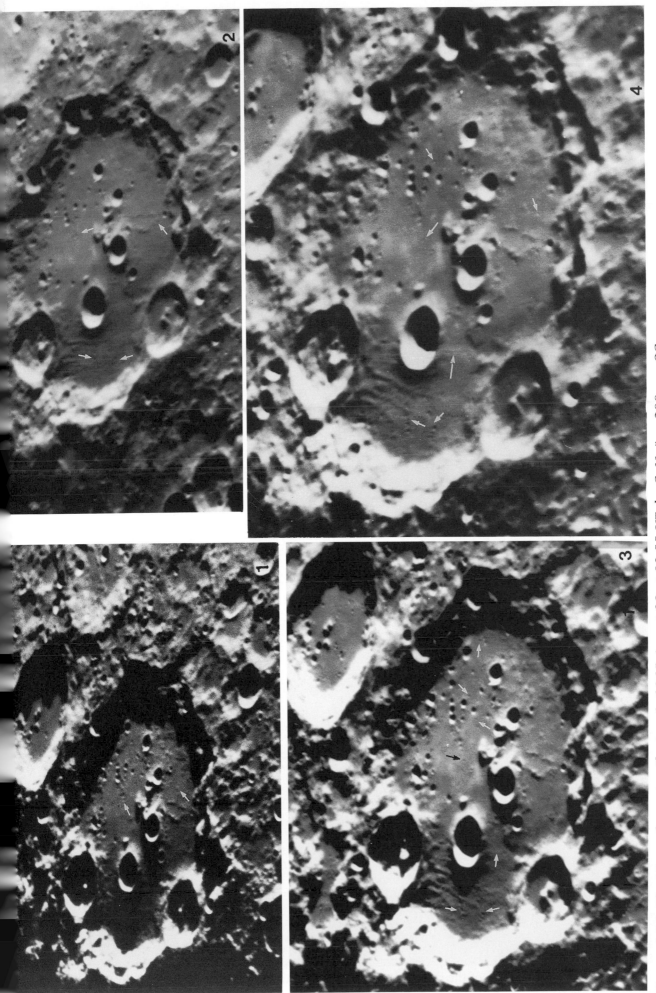

Figure 4.19. The Moon: the crater Clavius. 1. 1965 August 20 at 02.55 UT, by G. Neměc, 200 mm OG;
F/D = 80; 3 s. 2. 1987 August 26, by A. Behrend (La Chaux-de-Fond) with a 355 mm Schmidt–Cassegrain
(C14); 15 m focal length; 1 s on TP 2415 film. 3. 1992 August 21, by G. Thérin, with a 225 mm Takahashi
Schmidt–Cassegrain (F/D = 65); 1.5 s on TP 2415 film. 4. 1982 August 12 at 02.54 UT, by G. Viscardy
with his 520 mm Cassegrain, on TP 2415 film. Examination of the arrows is very instructive.

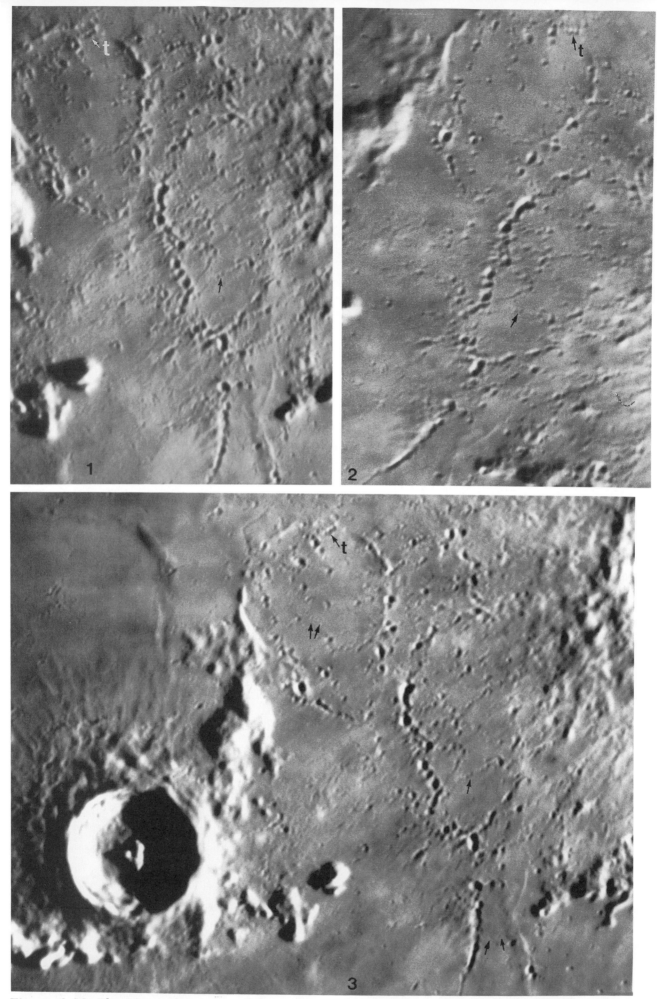

Figure 4.20. The Moon: the region of the crater Stadius and its famous triplet (t). *1. 1989 August 20, by B. Flach-Wilken (Wirges), with his 300 mm Schiefspiegler telescope; 2 s on TP 2415 film. 2. 1982 August 12 at 02.54 UT, by G. Viscardy with his 520 mm Cassegrain. 3. 1992 August 21,* by G. Thérin with his 225 mm Takahashi Schmidt–Cassegrain; F/D = 65; 1.5 s on TP 2415 film. The triplet (t) is resolved, into three adjoining craters, on the three photographs. The arrows denote some of the finest craterlets or rilles visible.

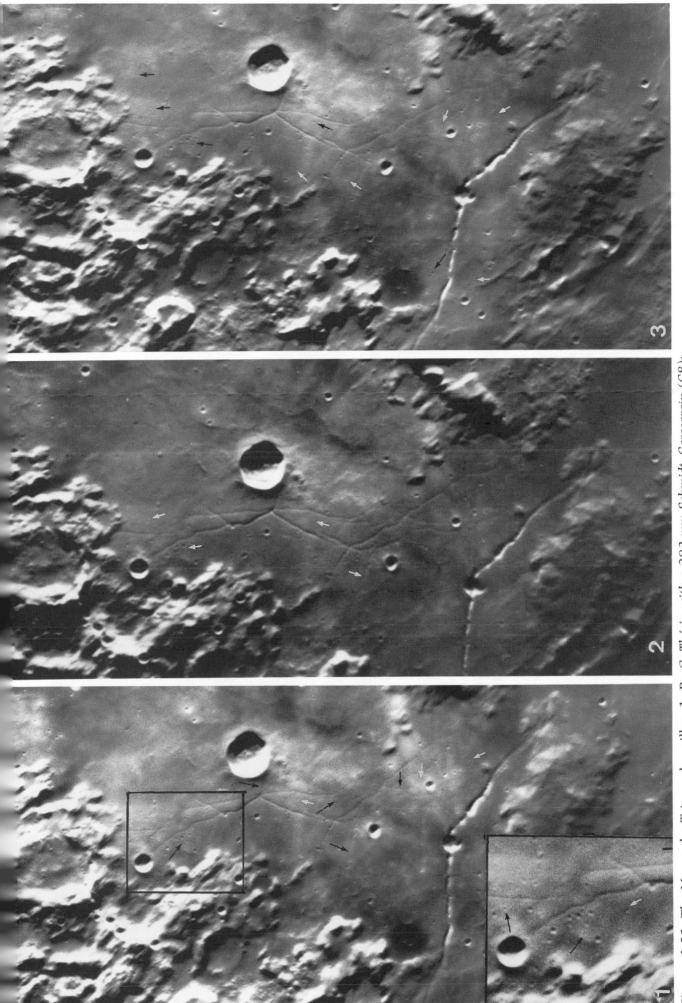

Figure 4.21. The Moon: the Triesnecker rilles. 1. By G. Thérin with a 203 mm Schmidt–Cassegrain (C8); F/D = 70; 1.2 s on TP 2415 film; developed in HC-110. Inset (1) shows tiny craterlets whose internal shadow measures only 0″.30. 2. 1986 August 27 at 03.00 UT, by B. Flach-Wilken (Wirges), with his 300 mm Schiefspiegler; 2.5 s on TP 2415 film. 3. No date, by C. Arsidi, with his 305 mm Cassegrain on TP 2415 film. Once again, examination of the small black and white arrows is very instructive.

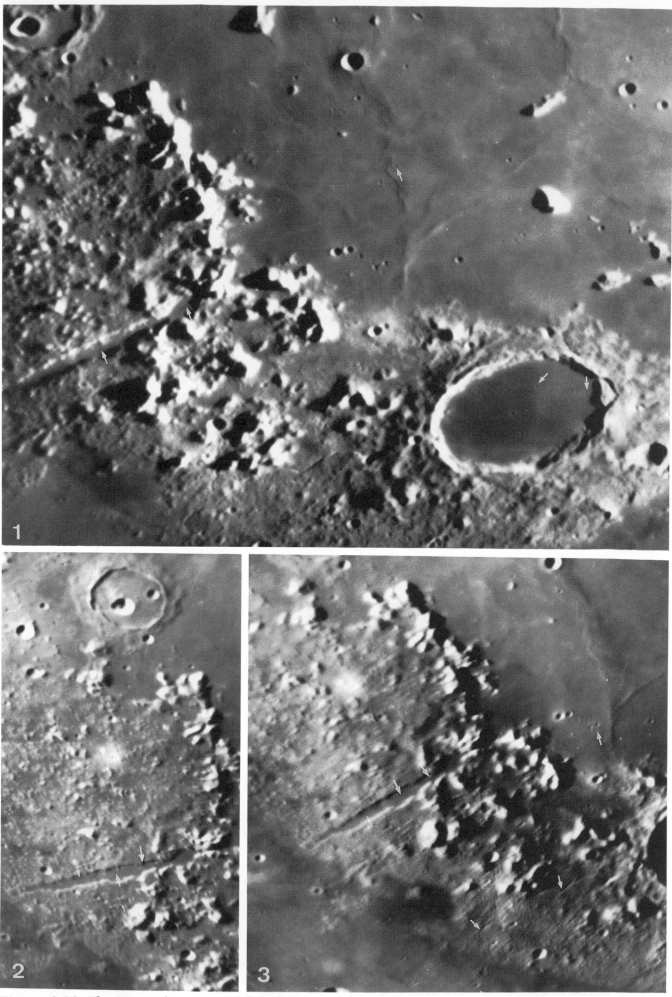

Figure 4.22. The Moon: the region of the Alpine Valley. 1. 1986 July 28 at 02.50 UT, by B. Flach-Wilken, with his 300 mm Schiefspiegler (20 m resultant focal length) on TP 2415 film. 2. No date, by D. Parker, with his 320 mm Newtonian. 3. 1981 January 14 at 18.10 UT, by the author with his C14; F/D = 60; 2 s on TP 2415 film. The cleft of the Alpine Valley (arrowed) is best seen in No. 3, owing to favourable illumination; the width of this cleft varies between 0".27 and 0".33. A difficult object to photograph well.

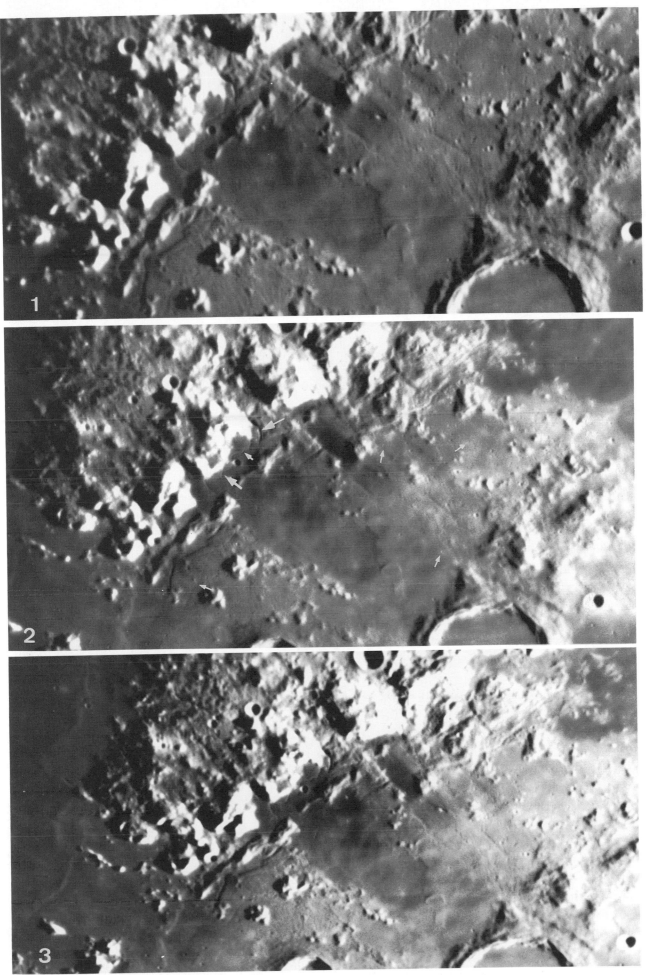

Figure 4.23. The Moon: the region of the Apennine Mountains. *1. 1986 July 28 at 02.50 UT, by B. Flach-Wilken (Wirges) with his 300 mm Schiefspiegler; 20 m focal length. 2. No date, by G. Thérin, with his C8; F/D = 70; 1.2 s on TP 2415 film. 3. No date, by C. Arsidi, with his 305 mm Cassegrain on TP 2415 film. A difficult region to photograph, owing to the strong contrasts in illumination.*

Figure 4.24. Lunar details. 1–3: The Straight Wall region. *1. 1965 August 19 at 02.10 UT, by G. Neměc, with his 200 mm OG. 2. 1984 September 10 at 03.08 UT, by G. Viscardy with his 520 mm Cassegrain. 3. 1992 April 11 at 19.44 UT, by Wolf Bickel (Bergisch Gladbach) with a 404 mm Newtonian and CCD camera; exposure of 6 × 0.08 s followed by image processing. The resolution obtained by Bickel, better than 0″.20, is incredible, but the* *field of view remains very small. 4. The crater Posidonius. 1982 November 5 at 02.26 UT, by G. Viscardy, with his 520 mm Cassegrain, on TP 2415 film. 5. The tortuous region of Walter, Aliacensis, Pernelius, etc. No date, by C. Arsidi, with his 305 mm Cassegrain; F/D = 45; 0.8 s on TP 2415 film. Very high resolution; 0″.3 or better. The arrows denote some of the finest craterlets or rilles visible.*

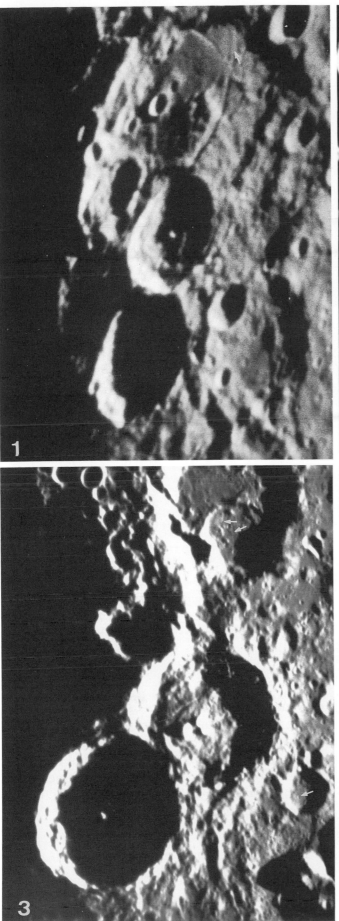

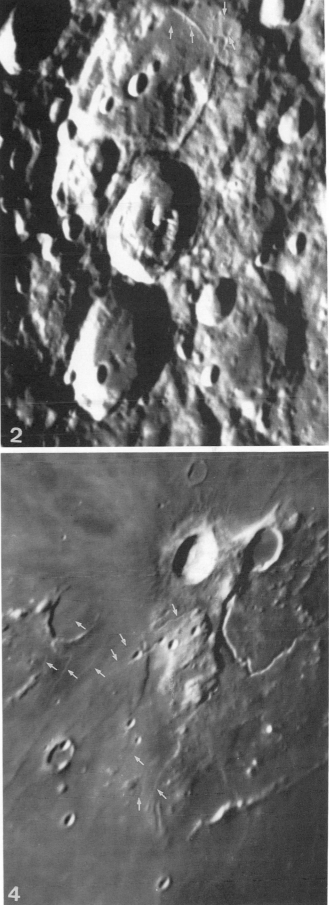

Figure 4.25. Lunar details. *1–2. The craters* Janssen, Fabricius, Metius, *etc. 1. 1964 November 23 at 00.00 UT, by G. Nemec, with his 200 mm OG; 18 m focal length; 6 s exposure (with correction for the drift in declination). 2. No date, by C. Arsidi, with his 305 mm Cassegrain. 3.* Theophilus, Cyrillus, Catharina. *No date, by C. Arsidi, with his 305 mm Cassegrain. Note the fine details in* Catharina. *4.* Aristarchus, Herodotus *and the rilles in the vicinity of* Prinz. *1981 February 4, by the author, with his C14; F/D = 60; 2 s on TP 2415 film. The arrows indicate the finest details (none of which are smaller than 0″.40).*

Figure 4.26. Lunar craters: *1. Petavius. No date, by S. Deconihout, with a 320 mm Cassegrain; F/D = 48; 1 s on TP 2415 film. 2. Gassendi, 1993, by G.* Thérin with his 225 mm Takahashi Schmidt–Cassegrain; F/D = 65; 1.4 s on TP 2415 film. The small arrows indicate the details of interest.

thanks to his exceptional ability and his thorough site-testing, successfully obtained lunar photographs equal to or better than most obtained at the greatest observatories, with a resolution of 0″.35 on craterlets and 0″.25 on rilles. Theoretically, a 20 cm C8, greatly hampered by a central obstruction of 0.34, should be incapable of recording details which have previously escaped the 60 cm refractor at Pic du Midi, the 91 cm refractor at Lick or the 2 m.50 reflector of Mount Wilson. In reality, however, it is possible. The German amateur Bernd Flach-Wilken obtained really beautiful lunar photographs with a 300 mm telescope. Many other amateurs have managed to take high quality photographs: Dany Cardoen (400 mm reflector), S. Deconihout (300 mm reflector), A. Behrend (355 mm reflector), L. Aerts (300 mm reflector), etc. The use of CCD receptors

will one day allow amateurs to go even further, owing to the shorter exposure times required and sophisticated image processing techniques. D. Parker has already photographed several of the craterlets on the floor of Plato by CCD camera with his 400 mm reflector. Figures 4.16–4.26 present a selection of the results obtained.

4.3.4 *The measurement and significance of the resolution and definition of lunar images*

Too often the terms 'resolving power' and 'definition' are confused. The former is characterised by the ability of the optical system to separate the two components of a close double star, or a pair of fine lines upon the Foucault testcard, whereas the latter refers to the chance of observing and photographing very small details, the visibility of which depends upon various factors, notably their shape and contrast (see pp. 14, 15 and Dragesco, 1969, 1992). It is generally thought that if a 200 mm objective has a theoretical resolving power of 0″.60, it will hardly allow lunar details smaller than this limit to be seen or photographed. This is evidently not the case. The 200 mm in question is able to render visible craterlets whose interior shadow is no smaller than 0″.30 and rilles no narrower than 0″.20.

The measurement of the smallest lunar details on a photograph presents real problems. In general, one measures the ratio between the smallest visible detail (often less than 1 mm) and that of the Moon (1–2 m, according to the focal length used and the degree of enlargement of the negative). Knowing this ratio and the average diameter of the Moon, it is possible to work out the real dimensions of the detail on the photograph, in kilometres or hundreds of metres, and then to translate these dimensions into arcseconds. However, the results are imprecise, and errors in excess of 30% can result. I have therefore adopted an alternative method: I have obtained several works, showing thousands of lunar photographs, obtained by the lunar Orbiters and some Apollo craft, at resolutions 10–100 times greater than those of the best Earth-based photographs. Many of these images from space (Schultz, 1976) are provided with a very precise scale in kilometres. With the aid of a good ×10 magnifying glass, equipped with a scale divided in tenths of millimetres, I can measure both the lunar detail and the image scale at the same time, with a maximum precision of 1/20 mm. By this means I obtain the dimensions of a lunar feature with high precision. For example, the fine rille which runs through the Alpine Valley measures – on average – 620 m in width (for a lunar diameter of 3476 km). At the time of calculating the angular dimensions, one can also take into account the variation in size of the Moon from apogee to perigee, provided that the date of the photo is known. On average, the rille described above subtends 0″.33: 0″.30 at apogee and 0″.35 at perigee. The distance of the feature from the centre of the disk must also be taken into account, and this will vary somewhat with libration. Most often, the rille is visible from the Earth under an angle of only about 0″.27. Unfortunately, the Orbiters did not photograph all the lunar features, and there are some gaps where the photos have no scale attached (the charts of Gutschewski et al., 1971). For the really fine details the famous ACIC charts can also be most useful (see below).

On the subject of resolution (or, to express it more correctly, definition) of a lunar photograph, I should add that it is a relative quantity: one measures the finest visible detail on an image (the width of a long, well-defined rille, for example) but the resolution found concerns only that detail. On a single lunar photograph, one could calculate a whole range of different resolving powers: a maximum definition of 0″.25 on a long rille close to the terminator, and only 1″.5 for low-contrast domes far from the terminator.

4.3.5 List of test objects of special interest in high or very high resolution photography

I have drawn up a list of 15 objects to observe and photograph which have interesting fine details (and for which I give the angle subtended at the mean lunar distance). See Fig. 4.27.

No. 1: Alphonsus (crater)
In this crater there is a whole series of fine rilles, of which just three can be photographed from Earth (Fig. 4.28:Al1). The largest (I) subtends an angle of 0″.32 on average, but is very irregular in width; it is very often photographed (Fig. 4.18). The second subtends no more than 0″.28; it is a difficult feature and seems to be the limit of a 200 mm telescope. The third, which is still finer (below 0″.25), seems to have evaded most photographers. The three should be observable visually with a good 300 mm telescope.

No. 2: Apennines (mountains)
In the familiar lunar Apennine 'mountain' chain towards the Palus Putredinus, several rilles within the scope of most apertures can be observed (Fig. 4.33:A). The finest of these, *Rima Hadley*, thin and tortuous, is often difficult to see, as it is lost in the shadows of the surrounding peaks; it is not easy to photograph, except under a very special angle of illumination. The long and wide *Rima Bradley* is within the scope of smaller instruments. The somewhat more tortuous rilles, named *Rimae Fresnel*, are equally well visible. The wider and less deep *Rimae Archimedes* need an oblique illumination to enable them to be seen.

No. 3: Catharina (crater)
In the interior of this old crater Catharina P can be seen, which is riddled with a multitude of small craterlets and shows a rille which is frequently photographed (Lick Observatory, Viscardy, Arsidi,

Figure 4.27. The Full Moon, *showing the locations of the 15 regions which the writer considers as suitable resolution tests: see p. 99 and Figs. 4.28–4.33.*

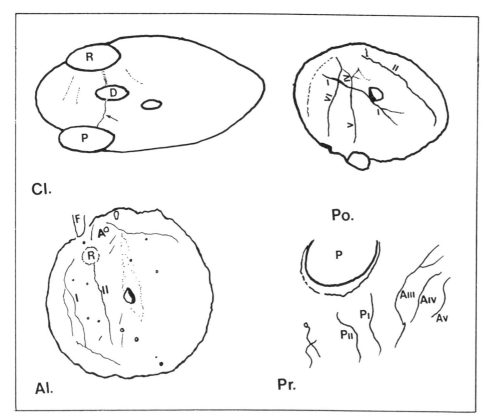

Figure 4.28. *Schematic views of the details to be photographed in the crater Clavius. CI: rille joining Clavius D (D) to Proctor (P) (arrowed). The crater Posidonius (Po): rimae I to VI, According to Orbiter imagery. The crater Alphonsus (Al): rimae I and II,* accessible to medium-sized telescopes. The numerous rimae to the north of the crater Prinz (P): P_1, P_2 and Aristarchus III, IV and V, accessible to amateur-sized instruments.

Thérin, etc.). One of the lunar Orbiters (photo No. 841) shows that its width does not exceed 0″.30 at any point. On ground-based photographs, one has the impression of there being several other fine rilles, but these are, in reality, only chance alignments of tiny craterlets.

No. 4: Clavius (crater)
Inside this enormous crater many craterlets can be seen, easily visible on the famous photograph taken with the 5 m Mount Palomar telescope. These details have been photographed by numerous astrophotographers, but the best image which I know of is that taken with the 1 m.50 telescope at Catalina, considered (in 1970) to be the best in the world (definition of 0″.15). The interest in Clavius resides in the problems posed by its rilles. In the past it was thought that there was a fine rille between Rutherford and Clavius D (Fig. 4.28:Cl). However, this rille is invisible in Orbiter photographs, with resolutions of some dozens of metres and under various conditions of lighting. A rille really exists between Clavius D and Porter (and strongly indents the north wall of Clavius D) but its width (measured on the excellent Orbiter photographs) is certainly less than 0″.15; it is particularly visible under a very high Sun. In my opinion, this rille is not perfectly clear except in one photograph: that of Catalina. It seems strongly enlarged by diffraction. It must be photographed again with precision, but is it accessible to an amateur telescope? There exist two other faults, very fine (less than 0″.10 in width), and rather shallow, to the east and west of Clavius, probably impossible to photograph from the Earth.

No. 5: Fracastorius (crater)
This crater, half-ruined, shows a host of interesting details but, again, remains incompletely known. It was photographed but once by Orbiter 4 (Photos 77-1, LPL 43302) and over only 4/5 of its surface. Documents for comparison are numerous but contradictory or inadequate. The drawings of selenographers are very deceptive. Fauth did not even see the great central crevasse, running east–west; Krieger first mistook it for a cliff; and the drawing of Wilkins and Whitaker (in Wilkins & Moore, 1961) seems quite fantastic to me. I should thank Dr John E. Westfall here, who provided me with a great deal of documentation on Fracastorius. Visible only in part in the famous Kuiper *Lunar Atlas* (91 cm Lick refractor and 60 cm refractor at Pic du Midi), the long east–west crevasse was afterwards

photographed by numerous telescopes (1 m.50 Catalina; 3 m Lick; 1 m Pic du Midi) and many amateurs such as G. Thérin. Other rilles are absent. That is why a photograph by G. Viscardy on 1982 November 5, 0234h UT is quite exceptional, as it shows a whole series of rilles (rimae) (Fig. 4.29:V). Some of these do not show up on other images taken on the same night (see Viscardy's *Atlas*, p. 418). The width of the finest rilles must be less than 0″.15. (I do not have an accurate measurement of their average width, owing to the absence of an Orbiter image with scale provided.) An ACIC chart (NASA) shows two or three rilles not imaged by Viscardy but also shows some rather unlikely details based on visual observations. The available Orbiter 4 photographs, which exhibit a higher resolution than the photographs of Viscardy, show only a few of the rimae photographed by the French astronomer. It therefore seems to me that Viscardy is the only person to have been able to record all the rilles in Fracastorius, as a result of special lighting conditions. Besides, this is not the only selenographic discovery by Viscardy; he has contributed in an important way to the 'Luna Incognita' programme of the ALPO. It is still possible to learn more about some poorly known features of the Moon (write to Dr J. E. Westfall, ALPO Executive Director, PO Box 16131, San Francisco, CA 94116, USA).

No. 6: Fra Mauro, Bonpland and Parry (craters)
These, again, are old craters, partially ruined. They are crossed by a whole series of rilles, generally very wide but not very deep, and not always easy to photograph. (See Fig. 4.33.) They are named *Rimae Parry Nos. I, II, III, IV and V.* The easiest is *Parry I*, which subtends 0″.65 in its widest part; *P IV* is smaller at 0″.55; and *P III* measures 0″.40. These rilles, although very broad, have been infrequently photographed.

No. 7: Gassendi (crater)
This crater is famous for the numerous crevasses (rilles, rimae) which cross its floor. Although not very narrow, these formations seem to be very hard to photograph. This is due to their obliquity and also to their low contrast, as they are not deep. The Orbiter photographs would lead one to think that the rilles should be visible with 20–25 cm telescopes. Figure 4.30:O shows the best-known features of the region, the principal rimae of which subtend: I, 0″.4; II, 0″.5–0″.6 or more (a very irregular rima); III, 0″.5; IV, 0″.5; V, 0″.35; VI, 0″.30. They have been easily and

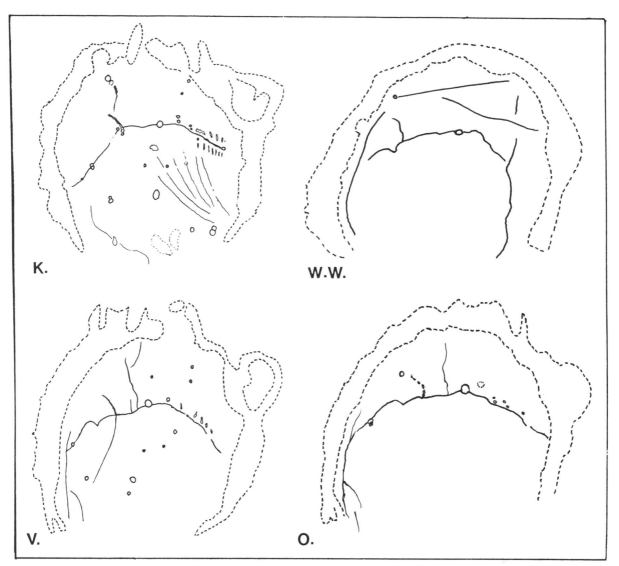

Figure 4.29. *The difficult problem posed by the crater* Fracastorius: *the chart published by Krieger (K) is more precise than that produced by Wilkins & Whitaker (W.W.); the chart that I drew up, after the photographs of G. Viscardy (V) on 1982 November 5, and finally, the diagram which the author was able to make according to existing Orbiter (O) data. The comparison of the two last diagrams shows that G. Viscardy could have discovered at least three new rilles (less than 0".10 in width).*

completely observed visually (Rükl, 1976). An effort should therefore be made to photograph them clearly. The drawings of Neissen and Fauth (Price, 1988), like that in Wilkins & Moore (1961), seem to me to be quite incredible, as they do not at all correspond to the images returned by the Orbiters. In general, all lunar drawings are extremely uncertain.

No. 8: Pitatus (crater)
This old crater is rich in rilles of various widths. Figure 4.33:P shows us what was seen by the Orbiters. Rima II is the easiest, as it reaches 1".2 across in its central portion. Rimae I and III are much harder, as they subtend between 0".22 and 0".45 (rather variable in width). Finally, the fine unnumbered rille seems to be at the limit of a 30 cm telescope, as it subtends only 0".15.

No. 9: Plato (crater)
This crater, the floor of which has been resurfaced, shows few details. At average resolution, only two craterlets can be photographed. In reality, there are at least six craterlets accessible to terrestrial telescopes. They subtend, respectively (in order of decreasing width): 1".5, 1".35, 1".2, 0".9 and 0".75 (if their shadows are distinctly visible, one can estimate

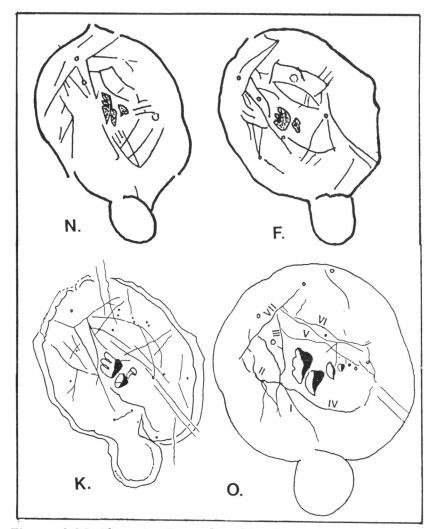

Figure 4.30. *The crater* Gassendi *and its rimae, first viewed by the selenographers Nasmyth (N), Fauth (F) and Krieger (K), and later according to Orbiter imagery (O). Note the discord between the maps based on drawings and reality!*

that the definition is better than 0″.7–0″.35) (Fig. 4.31:P). Plato being somewhat foreshortened by perspective, the visibility of its craterlets is rendered more difficult. It should be possible to photograph them all with amateur telescopes (some have, moreover, been photographed by B. Flach-Wilken with a 300 mm telescope, and have been recorded by the CCD receptor of D. Parker's 400 mm telescope). According to Wilkins (Wilkins & Moore, 1961), there are 16 craterlets to be seen visually through the 83 cm Meudon refractor. On the Orbiter photographs, at average resolution, they number more than 100.

No. 10: Posidonius (crater)
This is a spectacular crater, the floor of which is crossed in all directions by small crevasses. The majority are accessible to amateur instruments.

On Fig. 4.28:Po, I show, in particular, rimae I to IV (with a sort of marginal fault which I do not take into account). Rima I subtends 0″.50; IV and V measure between 0″.40 and 0″.30; and the final crooked extremity of IV is only about 0″.20 across. It should be possible to photograph them all (see Viscardy's *Atlas*, p. 312).

No. 11: Prinz (crater)
To the north of the partially flooded Prinz, and also of the crater Aristarchus, it is possible to see a whole series of fine, sinuous and not very deep crevasses of progressively decreasing width (Fig. 4.28:Pr). The largest are not hard to photograph. They subtend: *Rima Prinz I*, 0″.7; *Prinz II*, 0″.5–0″.6; *Rima Aristarchus III*, 0″.45; and *Rima Aristarchus IV*, 0″.32. There are two other sinuous rimae, which could be called *Prinz III and IV* (see

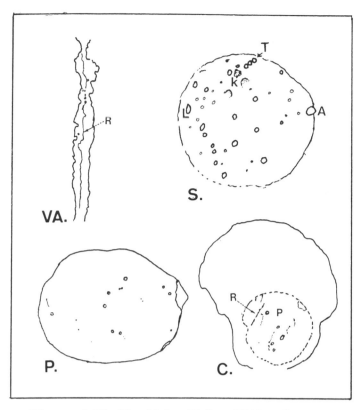

Figure 4.31. *The* Alpine Valley *(VA) and its central rille (R); the numerous craterlets of the crater* Stadius *(S) and the famous triplet (T); the crater* Plato *(P) and the dozen or so craterlets accessible to terrestrial telescopes; the crater* Catharina *(C) and its fine rille (R), in reality a crater-chain. Outlines traced from Orbiter photographs.*

Viscardy's *Atlas*, p. 338), which subtend about 0″.4. In reality, there is a whole series of other, much narrower crevasses (below 0″.25) which do not seem to have yet been photographed from Earth. Hill (1991) has drawn the region very well but he has introduced some distortions.

No. 12: Ptolemaeus (crater)
This great crater is especially celebrated for its 'saucers', or small depressions, visible only under oblique illumination but easily photographed. Ptolemaeus is notably riddled with small craterlets, the diameters of which subtend from 1″.1 to 0″.7 (internal shadows: 0″.55–0″.35, approximately). Most of the known photographs show only 30–40 craterlets. I have photographed about 50 of them myself with the 60 cm refractor of the Lowell Observatory (stopped down to 45 cm). Bernd Flach-Wilken has photographed almost as many (or maybe a few more) with his 300 mm Schiefspiegler. G. Thérin and, especially,

C. Arsidi have been able to photograph between 70 and 90. About 150 craterlets can be counted on one of the ACIC (NASA) maps, based upon the best terrestrial photographs and upon visual observations with the 50 and 60 cm refractors at Flagstaff (in fact, hundreds of them exist, as revealed by the Orbiters). The craterlets of Ptolemaeus therefore constitute a good test of high resolution lunar photography.

No. 13: Stadius (crater)
This very old crater is covered by a great number of craterlets, of which the famous 'triplet' is considered by Viscardy to be a good test for amateur telescopes. The triplet (Fig. 4.31:T) is composed of three interlinked craters, the diameters of which subtend between 1″.25 and 1″.55 but the spacings of which do not exceed 0″.27–0″.35. Thus, as Viscardy (1987) has remarked, if one can see that there are three craterlets on a photograph, without seeing their separations with certainty, the photograph exhibits a definition of about 0″.4. When the ramparts or interfaces are clearly visible, a definition of 0″.30 or more has been reached. It is a relatively easy test object, and yet it was not resolved by the Lick refractor (albeit on an old photograph only). Viscardy, Arsidi, Thérin, Flach-Wilken, Pic du Midi, Catalina, etc., have all easily resolved this test object (normal for a 300 mm, hard for a 200 mm aperture) (Fig. 4.20).

No. 14: Triesnecker (rille)
Everyone knows of the famous rilles or crevasses (rimae) which stretch between the craters Triesnecker, Hyginus and Rhaeticus. These formations are relatively wide but not very deep. They are almost all visible with amateur instruments. These rimae subtend between 0″.50 and 0″.40, some of them 0″.35 only (but are still visible on G. Thérin's photos taken with his 200 mm telescope as well as on those by B. Flach-Wilken and C. Arsidi). Numerous craters adorn this region, accessible to instruments of moderate power. They are between 1 and 3 km in size: the smallest subtend 0″.80 and their shadows 0″.30–0″.50. The selenographers Krieger, Molesworth and Wilkins (Price, 1988) drew a number of completely imaginary rilles. The Orbiters have impartially passed judgement upon their renditions (Fig. 4.32).

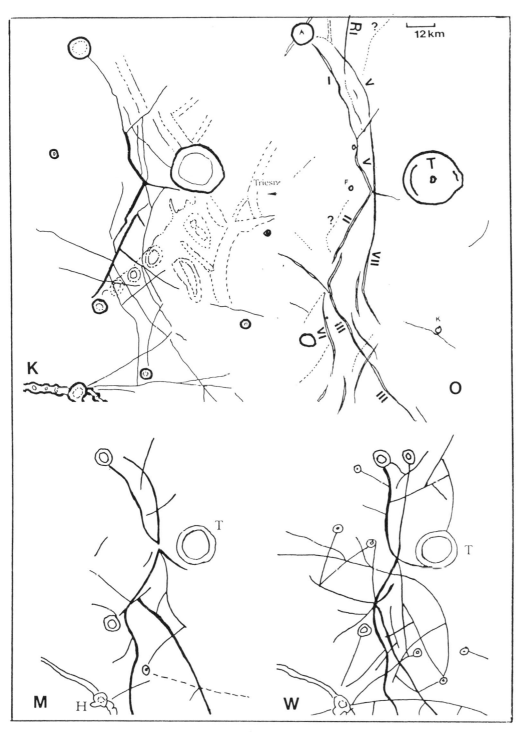

Figure 4.32. *Various interpretations of the* Triesnecker and Hyginus rilles, *after Krieger (K), Molesworth (M) and Wilkins (W), in comparison with reality: diagram O, drawn according to Orbiter imagery. The rimae I to VII can all be photographed with amateur instruments. The discord between the charts of these eminent selenographers and the real topography is, once again, striking.*

No. 15: The Valley of the Alps, or Alpine Valley (rille)
In the famous valley of the Alps the narrow, irregular rille which could not be photographed until recently is very well known. In theory it should not be too difficult a target, as it subtends 0".33. However, it is viewed obliquely and its apparent width does not exceed *c*. 0".27. Although it is easily visible in a 300 mm telescope, I am not aware of a single photograph with a 200–300 mm telescope which shows it clearly. It gave even Viscardy considerable trouble,

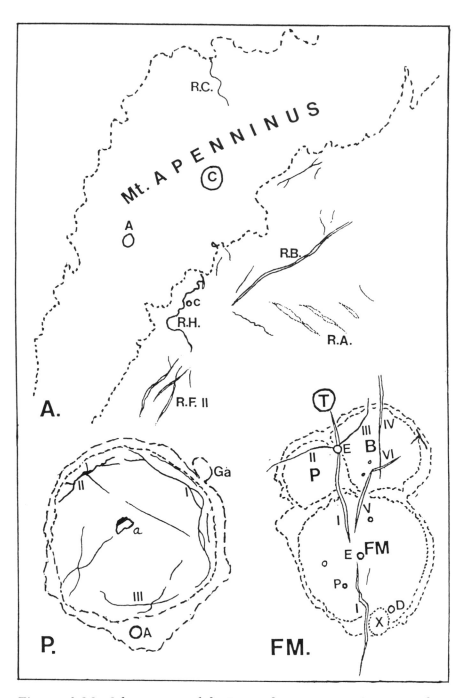

Figure 4.33. *Other regions of the Moon showing interesting topographic details. A.* The rimae of the Apennine Mountains: Rima Bradley *(R.B.),* Rima Hadley *(R.H.), as well as the* Rimae Fresnel *and* Archimedes, *all easily photographed. P. The crater* Pitatus *with rimae I and II, easily accessible, and others harder to photograph from the Earth. FM. The crater* Fra Mauro, *with the adjoining craters of* Parry *(P) and* Bonpland *(B), exhibiting well-marked rilles which have rarely been properly photographed.*

and I was not able to photograph more than three-fifths of its course with a Celestron 14 (Fig. 4.22). It is very well visible on the photographs taken with the 106 cm telescope of the Pic du Midi and the 150 cm at Catalina. It is therefore an interesting test for telescopes in the 300–400 mm range. D. Parker photographed it in part with a 300 mm telescope and C. Arsidi was able to do so as well; it can also be seen on the photographs of G. Thérin and B. Flach-Wilken.

4.3.6 Descriptive catalogue of lunar atlases based on photography

Lunar atlases by professionals

Lick Observatory Atlas of the Moon, by E. S. Holden, 1896–7. Photographs taken with the 91 cm refractor. Low resolution (focal length too short).

Photographisches Mond Atlas, by Weineck, Prague, 1899. Lick photographs further enlarged and completed with photographs from the Paris Observatory (see below). Low resolution.

Atlas Photographique de la lune, by Loewy and Puiseaux (Observatoire de Paris), 1896–1906. Photographs taken with the 60 cm Paris coudé refractor. Low resolution, due to the poor sensitivity of the emulsions of the day and to turbulence.

Atlas lunaire publié par la Société Belge d'Astronomie, 1899–1912. Simple reproduction of the principal photos of the *Atlas* of Loewy and Puiseaux.

Resumé del Atlas fotografico de la Luna des Observatorio Nacional de Paris, by Thort and Tanagona, 1922. Again, the photos of Loewy and Puiseaux, being the best of that epoch.

German reproduction of the Paris *Atlas*, published in Stüttgart.

Photographic Atlas of the Moon, by W. H. Pickering, *Annals of the Harvard College Observatory*, Volume 51, Cambridge, Mass., 1903. 88 plates taken with the help of a 300 mm telescope. Low resolution.

Photographic Atlas of the Moon, by S. Miyamoto and M. Matsui, Contributions from Kwasan Observatory, Kyoto, 1960. Photos obtained with a 300 mm telescope. Average resolution.

Photographic Lunar Atlas, by G. P. Kuiper, D. W. G. Arthur, E. Moore, Y. W. Tapscott and E. A. Whitaker, Chicago, 1960. 281 photographs, overenlarged, showing the main lunar views under different conditions of lighting. Contributions from the observatories of Lick, Mount Wilson, MacDonald, Pic du Midi and Yerkes. An enormous, very, very, heavy book. Usually average resolution, just occasionally very good (Mount Wilson and Pic du Midi).

Orthographic Atlas of the Moon, by D. W. G. Arthur and E. A. Whitaker, 1962. Reprint of part of the photos from the preceding atlas, with the lunar coordinates added.

Rectified Lunar Atlas, by E. A. Whitaker, G. A. Kuiper, W. H. Hartmann and L. H. Spradley, 1964. Images from the atlas of Kuiper *et al.*

(above), corrected for perspective by projection of the negatives onto a sphere. The original resolution was average, but has been very much altered by the rectification procedure.

Photographic Atlas of the Moon, by Z. Kopal, J. Klepesta and T. W. Rackham, Academic Press, 1965. Based on unpublished Pic du Midi 1961–2 lunar photographs (60 cm object glass), the same one as used 65 years earlier by Loewy and Puiseaux! Some 197 photos taken by the Manchester team. Unfortunately, the resolution is rather average for the most part.

Consolidated Lunar Atlas, by G. P. Kuiper, E. A. Whitaker, R. G. Strom, J. W. Fountain and S. M. Larson, Supplement Nos. 3 and 4, USAF *Photographic Atlas*, 1967. 192 actual photographic prints, from the best negatives obtained with the NASA 1 m.50 telescope of the observatory at Catalina (New Mexico), with another 34 obtained with the 1 m.50 Flagstaff Naval Observatory astrometric reflector. The prints are of excellent quality and the resolution is very high (reaching 0″.15). A work of great value.

New Photographic Atlas of the Moon, by Z. Kopal, Taplinger Publishing, New York 1971. A blend of photos taken with the 106 cm Pic du Midi telescope with photos taken from space; very variable resolution, from average to excellent (surpassing all previous works).

Lunar Astronautical Charts (LAC), published between 1960 and 1969. These were part of a series of publications of the ACIC of the USAF. The charts were drawn from the best photographs in existence (some being taken specially for this work by the Pic du Midi and Lowell observatories), completed by visual observations made at Flagstaff with the 50 and 60 cm refractors of the Lowell Observatory. These are very high resolution, well-presented, documents.

Lunar Orbiter Photographic Atlas of the Moon, prepared by D. E. Bowker and J. K. Hughes, NASA SP-206, 1971. Contains 675 superb photographs taken from space, with amazingly high resolution.

Atlas and Gazetteer of the Near Side of the Moon, by C. L. Gutschewski, D. C. Kinsler and E. A. Whitaker, NASA SP-241, 1971. 404 photos, taken by the Orbiters, accompanied by modern lunar nomenclature. Mediocre photographic reproductions, but with interesting details of techniques. An invaluable document for high resolution lunar photographers.

Moon Morphology, by P. Schultz, University of Texas Press, Austin, 1976. This is not an atlas as such, but a scientific study of lunar formations. Magnificent Orbiter photographs, provided with a scale which allows the exact actual dimensions of the features to be obtained. Extremely valuable for determining the definition of amateur photographs.

Popular lunar atlases

Atlas de la Lune, by V. de Callatay, Paris, 1962. Numerous Pic du Midi photographs and some from American observatories. Very old documents, poorly printed, deceptive resolution.

Pictorial Guide to the Moon, by D. Alter, T. V. Cowell, New York; *Lunar Atlas*, by D. Alter, 1964 (also 1968), Dover Publications, New York. These two atlases are based primarily on old photographs, obtained at the Lick Observatory and at some other American observatories: 154 photos for the *Lunar Atlas*, extremely poor reproduction and mediocre resolution.

The Times Atlas of the Moon, by H. A. G. Lewis, New York and London, 1969. Maps drawn from the best American LAC and ACIC documents. Beautiful quality and very high resolution.

Hamlyn Atlas of the Moon, by A. Rükl and T. W. Rackham, Hamlyn, 1991. Maps and 50 'photos' from the *Consolidated Atlas*. Unfortunately, the resolution of the chart section is low (with the exception of the rimae, which were checked with the Orbiter photos). The selected photographs are not the best and the reproductions are actually *drawings* of them (a bizarre practice!!).

Lunar atlases by amateurs

Amateur astronomers' Photographic Lunar Atlas, by H. Hatfield, Lutterworth Press, London, 1968. 111 photographs of the Moon, taken with a 300 mm reflector. Very complete but low resolution, owing to turbulence. (As a guide for beginners, it serves its purpose well.)

Atlas-guide photographique de la lune, by G. Viscardy, Association Franco-Monégasque d'Astronomie, 1987, Masson editions, Paris and distributed by Willmann-Bell in the USA, 451 pp. 246 high resolution photographs, taken with 310 and 520 mm telescopes. Very complete (even for the poorly known regions of the Moon), with a detailed nomenclature and a very interesting text. New chapters on

selenographic research by amateurs. A monumental work.

Photographic Guide to the Moon, by M. Shirao and S. Sato, Rippu Shobo, Tokyo, 1987, 151 pp. (in Japanese). Numerous photographs, but low resolution (160–200 mm telescopes).

Berliner Mondatlas, by A. Voigt and H. Giebler, Wilhelm Foerster Observatory, W. Berlin, 1989. Reprint of an old 1960s publication. Original photographs obtained with the 30 cm Bamberg refractor. Very poor resolution. Of little interest.

Fotografischer Mondatlas, by W. Schwinge, 1983, Ed. J. A. Barth, Leipzig. Original photos obtained with a 110 mm refractor and a 300 mm reflector. Impossibly low resolution. Hardly credible that such a work could be published. Only interest: detailed nomenclature and very low price.

There are a number of charts or lunar atlases based solely on drawings but they are not included in the above list.

4.3.7 Final remarks

It may appear odd that, over so many years, some of the world's most famous telescopes have failed to produce anything except relatively disappointing results in the field of lunar photography. There are various reasons for this: the objectives or mirrors of the most powerful telescopes have not been made to the highest precision; their great apertures make them very susceptible to turbulence, all the more as their famous sites enjoy no better than average seeing (in stellar photography, a resolution of $1''$ is often considered satisfactory); finally, the photographic methods adopted are, generally, inappropriate (focal ratio F/D too low, emulsions too grainy, etc.). That is why the best lunar photographs, obtained at the largest observatories, are the products of the refractors and reflectors of from 60 cm to 1 m.50 diameter. To date, the 106 cm telescope of the Pic du Midi has seemed best suited to this type of work. It must also be said that the professionals have abandoned lunar photography, as the Ranger, Orbiter and Apollo programmes (together with the recent Clementine missions) furnished thousands of photographs having a resolution 10–100 times higher than had been achieved earlier. The field is therefore open to amateurs, and, profiting from the benefit of more sensitive emulsions and improved photographic techniques, they have been able to show, thanks to their tenacity, that 200–300 mm telescopes are able to yield images just as good as the best 'ancient' photographs of the great observatories. The limit of definition of 0″.15, obtained at the Catalina and Pic du Midi observatories, is within reach of an excellent 400 mm telescope, given perfect seeing. Certainly, the amateur will be left far behind by the Hubble Space Telescope (revised and corrected) or by telescopes with adaptive optics. It is likely that, even today, the Pic du Midi team who work with the 2 m telescope on lunar geology has already obtained photographs with a definition exceeding 0″.10. However, amateurs need not fear competition. Lunar photography at high resolution will remain their field. It is the ideal test for all who want to reach the photographic limits of their instrument. The CCD receptors of the future, of large diameter, will allow for further advances.

I reproduce, intentionally, some lunar drawings based on visual observations made by some of the best selenographers of the past: Nasmyth, Fauth, Krieger, Wilkins, Molesworth, etc. Their charts are erroneous, and haunted by imaginary details (the Orbiters are merciless!).

4.4 High resolution planetary photography

4.4.1 Historical introduction

The field of planetary photography is one of the most difficult because the planets have small disks and their features are usually not very apparent (low contrast). In addition, we have to try to photograph details as small as 0″.20.

Since photography of the planets has always been a delicate matter, few observatories have really been able to master it. Atmospheric turbulence remains the main obstacle – an obstacle which, of course, increases with the diameter of the instrument used. It is therefore not the largest telescopes which have brought the best crop of results.

Thus, as Dollfus (1961) writes, irrespective of the diameter of the telescope used, the finest planetary details, photographed at the majority of the world's observatories (from the 38 cm OG at Pic du Midi up to the 5 m Mount Palomar telescope), vary around the 0″.40 mark. Over many years, professional astronomers did not have access to sensitive emulsions appropriate for planetary photography; hence the mediocrity of many such photographs published between 1946 and 1970. It will suffice to leaf through the books by Slipher (1962, 1964) to see how much the 60 cm Lowell refractor (most often diaphragmed to 45, even 32 cm) – in spite of the thousands of

photographs obtained at Flagstaff – was severely handicapped by poor seeing (all the more as the exposure times reached 2–3 s for Mars and 6–8 s for Jupiter). The average resolution of these photographs is rarely better than 0″.6. It therefore seems certain, today, that the site at Flagstaff is not as wonderful as its founders liked to think. Percival Lowell himself wrote to E. M. Antoniadi in 1909, concerning the 83 cm refractor at Meudon: '. . . remember that you will have to diaphragm it down to get the finest possible details. Even here, we find 12 or 18 inches the best size' (McKim, 1993). Happily, Antoniadi did not follow Lowell's advice. Slipher himself (1962) admitted that the image of the planet Mars was constantly moving by more than 1″. He was to try, throughout his career at Flagstaff, to reduce the exposure time.

One could make the same comments about the photos by Finsen, in South Africa, which remain mediocre, despite compositing. Finsen worked with 16 mm Kodakchrome (therefore with too low a focal ratio) and had to diaphragm his 65 cm objective to only 34 cm!

It was, therefore, B. Lyot and his co-workers H. Camichel and M. Gentili who really initiated high quality planetary photography, with the 60 cm refractor of the Pic du Midi observatory. Nevertheless, their fine photos rarely revealed details smaller than 0″.40. The 106 cm telescope later installed at the Pic allowed P. Guérin, and later Camichel and Boyer, to obtain what were again among the best known results (1966–80). This, however, was by no means easy. I was at the Pic in 1973 and 1975 and was able to work with C. Boyer and the new planetary camera, which allowed enormous images ($F/D = 92$) to be obtained on Ilford Pan F emulsion, developed in D19. In spite of energetic development, the pictures exhibited a terribly low contrast. However, this did not prevent Boyer from securing, on 1975 October 2, 2240h UT, one of the most remarkable photographs of Jupiter ever obtained, showing details smaller than 0″.20. It was not until the introduction of TP 2415 film (1980) and, above all, after having replaced the photographic film by the CCD camera, that the 106 cm telescope was able to demonstrate its full capabilities (in the hands of MM. Lecacheux, Colas, Buil, Thouvenot, etc.).

Earlier, E. J. Reese, H. G. Solberg and B. Smith had been able to show that the 'small' 300 mm telescope at New Mexico (Tortuga Mountain Observatory) was capable of providing excellent high resolution planetary photographs. (The later installation of a 60 cm telescope has not perceptibly improved the resolution obtained.)

At Mount Teide Observatory (Tenerife), our Spanish amateur friends Gomez and Tomas obtained magnificent Jupiter photographs with the 40 cm vacuum solar telescope. The fine images of Teide have proved, more than once, that a 400 mm telescope is usually large enough to obtain excellent planetary photographs.

When NASA decided to initiate the International Planetary Patrol programme, they chose to install 600 mm telescopes, distributed all around the globe, allowing round the clock photography of the planets Mars and Jupiter, with acceptable resolution (although very variable according to location: from about 0".35 to 0".80).

About 1970, the Catalina Mountain Observatory complex attached to the University of Arizona, was given a 1 m.50 Cassegrain telescope by NASA, for the use of the Lunar and Planetary Laboratory, directed by G. Kuiper. The telescope (of excellent quality) was installed at the Catalina site, at 2600 m (8300 ft) altitude, and became the great competitor of the Pic du Midi's 106 cm. G.Kuiper was aided by numerous co-workers (S. Larson, J. W. Fountain, R. B. Minton, R. le Pool, E. A. Whitaker, etc.), and the publications of the Lunar and Planetary Laboratory of the University of Arizona have demonstrated what the 1 m.50 telescope could do: resolution better than 0".15 on the Moon, better than 0".20 on Jupiter. Kuiper (1972) calculated that between 1920 and 1971 there had been obtained, in all the world's observatories, more than one million lunar and planetary photographs, of very variable quality. Although, from 1890, double stars could be resolved visually at 0".10, it would take more than 70 years for this performance to be equalled even once in the field of planetary photography. Kuiper (1972) thought, in reality, that the 1 m.50 Catalina telescope allowed all the details seen visually to be photographed, i.e. details smaller than 0".15 (whereas at the MacDonald Observatory, in 1948, the photographic resolution was four times lower than the visual resolution). In fact, even the great Kuiper exaggerated and always made the mistake of confusing the resolving power for double stars with the definition of an extended object. The 1 m.50 telescope has a theoretical resolving power of 0".08, from Dawes' formula. That is to say, it should be possible to photograph isolated details between 0".04 and 0".01 (see the explanations on p. 14). We are still far from these limits.

The resolution has been improved by the use of CCD receptors, but the true capabilities of a 1 m.50 telescope have still not been realised. Only the Hubble Space Telescope will be able to show us planetary details 0".01 in size.

4.4.2 Instrumental and atmospheric considerations

High resolution planetary photography is rarely open to the amateur. More so than for lunar photography, it requires very stable images and therefore very special sites. I shall not return here to the problems posed by external turbulence, either local or instrumental (see pp. 3–8). A very simple test allows us to check, as for the Moon, whether high resolution planetary photography is possible or not: a planet is observed, with a high power (×200–300) eyepiece, equipped with a fine grid or reticule, dimly illuminated. Matters are arranged so that the planet is placed tangentially to two perpendicular wires (or rulings), and then its behaviour observed. The planet will not stay still but will be displaced in all directions (owing to turbulence), by an amount which is easily determined if the apparent diameter of the planet in arcseconds is known. If during 3 s of time the image moves by one-fifth of the diameter of the planet, i.e. by 3–8", it is useless to try to photograph it. To obtain a photograph worthy of the 'high resolution' label, the planet must remain almost motionless throughout 2–3 s of time, which, of course, is hardly ever the case. It is therefore clear how planetary photography at high resolution is difficult and also how exceptional the site of Okinawa is, from which Isao Miyazaki records details as small as 0".25 upon Jupiter, even with exposure times of 4 s! The site of D. Parker at Coral Gables (Florida) is almost as good. G. Viscardy readily confirms that it is impossible to use long exposures anywhere in France, even in a selected site as good as his. I myself considered at Cotonou (Benin) and elsewhere in Africa, how such tropical or equatorial regions, hot and humid, with low thermal changes in 24 h, could constitute especially favourable sites for high resolution (for more details about this, see pp. 5–7). On the other hand, and contrary to the opinion of several specialists, I do not think highly of the advantage of high altitude for obtaining good images, at least not unless one goes up to at least 5000–6000 m. Altitude certainly plays a role in the Continental regions (Pic du Midi, Catalina), which have contrasting climates and constant atmospheric instability.

In the Equatorial countries (from Cotonou to

Brazzaville and from Makokou to Boscha, Java) excellent images can be obtained at sea level; on the other hand, transparency is less favourable, necessitating exposure times twice as long as under our skies, and being unfavourable for UV photography. Perhaps still better images could be obtained on the summit of Mount Kilimanjaro, or Mounts Cameroun or Ruwenzori, but I strongly doubt it, judging from the force of the winds and the very large changes in temperature during 24 h which prevail on these summits. In what are called temperate countries (notably France, Great Britain, Germany and the eastern United States) it is the turbulence, always present, which leads to the instability of the images, from which follows the need to keep the exposures as short as possible: 1–3 s. This is what we try to do, with more or less success. We would be able significantly to improve our results if we had the advantage of a receptor at least ten times more sensitive than our famous miracle film TP 2415 (when overdeveloped). In other respects, it is curious to record that the professionals do not seem to have discovered that it is not really necessary to use very large telescopes for high resolution planetary photography. We now know, though admittedly since only recently, that a 400 mm telescope of high quality suffices to record details down to 0″.25, which is at the level of resolution of the photos obtained with telescopes of 1–2 m in size. The figure of 0″.10 that G. Kuiper (1972) dreamed of still remains hard to attain and requires instruments of 1 m aperture or larger.

So, with a 400 mm telescope under good seeing conditions at Okinawa or at Coral Gables, photos are now obtained (upon film or CCD receptor) which rank with those secured at Pic du Midi or Catalina and are usually much better than those which were obtained a few years ago at Flagstaff or at the NASA stations. However, the sites of Okinawa and Coral Gables are exceptional. In conclusion, at an ordinary site there is no special reason for exceeding 300 mm diameter. Apertures of 250 to 400 mm should be used to suit the quality of the site; the 520 mm at St. Martin-de-Peille is certainly oversized.

The refractor, in spite of the popular revival which it is enjoying (thanks to apochromatic, short-focus objectives), does not seem to me especially suitable for planetary photography. The diameter of 180 mm (already expensive and cumbersome) is insufficient, and, in my view, inferior in resolution to a 250 mm mirror. I would not insist on this opinion if I had not worked for

some 30 nights with the Alvan Clark refractor at Lowell Observatory: in spite of its diameter of 60 cm, the planetary photos which I was able to obtain, in September–October 1988, were largely inferior to those which I had been able to take at Cotonou with my 355–mm Schmidt–Cassegrain (C14); the Flagstaff refractor should be diaphragmed down to between 32 cm and 45 cm (just as Percival Lowell did, 90 years ago!), either because of possible zonal defects of the objective or because the latter sizes are the limit for that site. I was able to obtain only once a resolution of 0″.30 on the Moon, but all my 2000 negatives of Mars and Jupiter were mediocre. Besides, it will suffice to leaf through Slipher's two books (1962, 1964) to realise how the famous planetographer had to be content with little, and that in spite of compositing he never obtained more than a few good photos each opposition (resolution about 0″.40). He published many images which would be considered unacceptably poor today.

The reflector, then, would seem to assert itself. But high resolution demands near-perfect optics, and the Newtonian is the only type of reflector which can be made, without too much trouble, to a quasi-perfect standard ($\lambda/20$ or better). It is surely not by chance that the best planetary photographers of our time work with $F/D = 6$ Newtonians. The Cassegrain is often also used, as it permits a high focal ratio to be obtained, up to $F/D = 60$, in certain cases. However, the production of the small hyperbolic secondary remains difficult. On the other hand, the supporters of the Cassegrain generally opt for a primary with a low focal ratio ($F/D = 4$), which makes the production of the mirror difficult and leads to a large central obstruction. I shall not go into the details of the Schmidt–Cassegrain again here; see p. 22. Theoretically, their optical specification does not especially commend them to high resolution planetary work. That said, the Celestron 14 (355 mm) has allowed me to obtain some hundreds of very fair Jupiter photographs during my stay at Cotonou (Benin). Many of these photos are inadequate in terms of today's criteria (Fig. 4.44:1). The tri-Schiefspiegler arrangement, with three mirrors, devoid of central obstruction but hard to make correctly, has given very good results in the hands of B. Flach-Wilken and T. Platt.

However, a good set of 250–300 mm optics cannot be conveniently used if the mounting is not equally good: solid, heavy, perfectly aligned and, above all, provided with a perfect drive, continuously checked at the eyepiece with a graticule. The planet should remain central, upon the graticule, for at least 4–5 s. The precision required is 0″.30, which is not very easy to attain, and few amateur telescopes permit it. At Cotonou the resolution of my planetary photographs was limited by the drive of my Schmidt–Cassegrain, the performance of which was irregular. If Isao Miyazaki dares to expose for as long as 4.5 s and obtains, even then, a Jupiter negative having a resolution better than 0″.30, it is due not only to a near-zero turbulence but also to a near-perfect drive (worm-wheel 396 mm in diameter, cut to a very high precision). Note that it is possible to drive amateur telescopes with a Texereau-type drive (see Verseau, 1986, and pp. 28, 29), which works better than most commercial drives.

Finally, as with lunar photography, it is very important to check the collimation of the telescope before each photographic session, and to keep an eye on the temperature within the telescope tube, to see that it is the same as that of the surrounding air (a precision mercury thermometer should be permanently fixed inside the tube at the level of the main mirror). If the cooling of the telescope interior to ambient temperature is very slow, especially if the telescope is sealed with an optical window or by an object-glass, then it should occasionally be ventilated by a fan. Good use is made of this method at Pic du Midi and at Catalina. Takahashi offer, as an optional extra, on their latest Schmidt–Cassegrain SCT 225 a very efficient fan. On the latest Mewlon 250 Dall–Kirkham, Cassegrain by the same manufacturer, the lower end of the tube beneath the main mirror may be removed, so that the surrounding air can circulate freely both behind and around the mirror.

4.4.3 Optical and photographic considerations

Opinions about the degree of amplification of the primary image in planetary photography are controversial. Most astrophotographers choose focal ratios F/D ranging from 70 to 120 (70–80 for Jupiter, 80–100 for Mars and Saturn). Parker and Miyazaki are exceptional in adopting very large amplifications: $F/D = 100$–200, with an average of about 140 (which means a final effective focal length of 56 m). These high amplifications allow them to obtain large images on their negatives, leading to a total absence of grain on the final prints. Nonetheless, such great focal lengths entail three inconveniences: (1) a significant increase in the exposure time (which can reach 20 s with Saturn); (2) an extremely

small field (which entails an auxiliary telescope – a high power finder, ×20–25); (3) a very blurred image on the glass focusing screen (necessitating a low power eyelens ×2–3 instead of ×5–6). As far as I am concerned, I have little interest in such high amplifications. In 1986 July, with the 106 cm telescope of Pic du Midi, I obtained one of the best photographs of Mars taken up to that time (Fig. 4.38); the image on the negative, however, measured hardly 5 mm across ($F/D =$ about 48). The resolution does not seem to be limited by the grain of the film (even if the latter is perceptible). I thus continue to think that with telescopes between 250 and 400 mm in size a focal ratio F/D between 60 and 100 should be sufficient.

The final amplification can be obtained in various ways. At the time of my stay at the Pic du Midi in 1975, C. Boyer tried out the new planetary camera designed by himself and Camichel, which used two negative Barlow Clavé ×3 doublets (50 mm diameter), allowing a focal ratio F/D of 92 to be obtained (in order not to be handicapped by the obtrusive grain of Ilford Pan F film, developed in D19). At more than 96 m focal length, the images on the negatives became enormous and focusing was not very easy in only average seeing.

When one only has a Newtonian working at $F/D = 5$–6, the primary image should be enlarged ×10–25. One is therefore forced to use high power eyepieces, and to choose them carefully. Miyazaki sometimes uses a special Pentax XP eyepiece, with a focal length of 3.8 mm. One can also try the short-focus Nägler eyepieces, the various Vixen LV high resolution eyepieces (4 and 5 mm focal length) or the projection eyepieces by Takahashi. A good solution consists of using a microscope objective which, for a projection distance of 160 mm, gives perfect images. The amplification is equal to its stated magnification (engraved on the side): a ×10 objective converts a system working at $F/D = 6$ to one working at $F/D = 60$, and an objective of ×16 to one of $F/D = 96$, etc. It is important to choose a quality objective (Zeiss, Leitz, Nikon) provided with a flat field and a high numerical aperture. An apochromatically corrected objective is not necessary for small magnifications. The front lens of the objective should be facing the side of the focal plane (or primary) image to be enlarged.

When working with a Cassegrain or Schmidt–Cassegrain, or with a refractor having a focal ratio of $F/D = 10$–12, one can choose between a microscope objective of ×6–10, or two Barlow lenses, coupled together or separated by a short distance. By varying the projection distance one should be able to attain a final focal ratio of between 60 and 80. One can also use a high power eyepiece (8–12 mm focus) with a short projection distance, but it would be better to choose a so-called 'projection lens'; see p. 32.

For planetary photography one can use an image format below 24×36 mm (24×24 mm or 18×24 mm) in order to economise on film. In general, a 24×36 mm reflex camera is used, provided with a clear focusing screen with graticule, and a viewing lens the magnification of which should be inversely proportional to the focal ratio chosen. As Dobbins *et al.* (1988) state, it must not be forgotten that the focal image is enlarged twice: once by the eyepiece and once by the viewing lens. Assume that we are working at $F/D = 150$, with a 400 mm telescope. The resultant focal length will be 60 m. If this image is observed with a magnifying lens of 50 mm focus (magnification ×5), as is usual with a reflex camera, the final magnification will be ×1200. Thus, the planets appear enormous and blurred, and focusing will be very difficult. That is why D. Parker has made a viewing lens of only ×2 magnification, with the help of lenses purchased from Jäeger (USA). The optimum magnifications are ×5–6 for $F/D = 50$–60; ×4 for $F/D = 80$–100; ×3 for $F/D = 120$–180.

In every way, viewed at the high power obtained by the double enlargement of the focal plane image (the first for obtaining the focal ratio desired, the second for observing the image upon the focusing screen), planetary images remain very blurred and the surface details are poorly seen. One must therefore focus either on the satellites of Jupiter, the limb of Mars (particularly on the polar cap) or the outline of Saturn's rings. When the focal length has a tendency to vary, through the effects of turbulence, one must check the focus frequently (which is why focusing on a star should be avoided). With a focal ratio of $F/D = 80$, a tolerable focus can be obtained with the help of an Olympus Varimagni viewer, set to its lowest magnification. The ideal would be to have a zoom-type viewing lens, allowing magnifications from ×2 to ×6, and therefore compatible with focal ratios from about 50 to 180. The use of a 24×36 mm reflex camera presents several problems: wastage of film (a planet occupies no more than a few millimetres on a piece of 24×36 mm film); vibrations produced by the flipping of the reflex mirror and the clicking of the shutter (which necessitates a full aperture

shutter, or the 'hat-trick' method; see p. 39); and the impossibility of controlling the state of the turbulence at the moment of exposure. In photography there can therefore be problems even in favourable moments, those happy conjunctions between good seeing and perfect alignment. Parker has thus tried to improve his chances with his 'seeing monitor' (see p. 36), while I have devised two convenient pieces of original photographic apparatus allowing me to make use of the best moments (see pp. 36, 37 and Figs. 3.2 and 3.3). If one uses a commercial reflex camera, it is important to have a motor drive to advance the film and prime the shutter, so as not to interrupt the perfect meshing of the worm and wheel. A film dater is also a useful accessory, as every planetary photograph is a unique document which loses all its value if the exact date and time (in UT!) are omitted.

Like lunar photography, planetary photography requires the use of a film having apparently contradictory characteristics: good general sensitivity, high contrast, very fine grain and high resolving power. We know that such a film exists and that it has overrun the entire field of high resolution photography: the famous Kodak Technical Pan 2415. It must be purchased in large 45 m spools, as the consumption of film in planetary photography is high (some hundreds, in truth thousands, of negatives when Mars and Jupiter are observable at the same time of year).

The brilliance of the planets (more exactly, their albedos) are very variable. If we take Venus as unity, Mars is 22 times, Jupiter is 77 times and Saturn 260 times less bright (see p. 40). The focal ratio F/D cannot be varied by much, so the maximum sensitivity of the film must be utilised, in order to shorten the exposure time as much as possible.

Unfortunately opinions differ widely concerning the development techniques for planetary negatives. Traditionally, D19 developer is generally used, as the films of the past (Pan F, Tri X, Adox KB 17) were less contrasty. One can still use D19 with TP 2415 or, alternatively, Dektol (3–4 min at 20 °C), which allows increased contrast and sensitivity to be obtained. Dobbins *et al.* (1988) have for a long time insisted on the necessity of using a developer capable of bringing out all possible gradations in tone. I am in complete disagreement with this. The photography of the planets cannot be compared with landscape photography (see p. 49). On Jupiter and Saturn the atmospheric details which one seeks to photograph show a low overall contrast, and if

113

Mars shows better defined features, the small details are, in the same manner, less contrasty, owing to diffraction (above all with 200–300 mm telescopes, where one is working at the limit of resolution). I therefore think that planetary photographs must be developed to a γ rating between 1.50 and 2.50 or to a CI ranging from 1 to 1.80, whereas Parker considers that the best contrasts are CI = 0.6 for Mars and 0.7–0.8 for Jupiter and Saturn. In fact I am ignoring the figures published by Dobbins *et al.* (1988) on p. 177 of their book concerning Contrast Indices for TP 2415 film developed in Rodinal, as they seem to me to be too low and not to correspond with my own results (see p. 44 and Tables 3.9 and 3.12). I therefore think that Parker and Miyazaki are working at Contrast Indices of at least 1.25. They thus obtain excellent results by the use of Rodinal (dilution 1:50 for 14 min at 20 °C or 1:100 for 12–14 min) as the developer. Parker thinks that Rodinal is advantageous because of its excellent acutance and its compensating effect (bringing out detail in the shadows and underexposed regions) but he recognises the light loss in sensitivity, with regard to less forcing development conditions; see p. 49 and Tables 3.9 and 3.12. That is why, it seems, Parker and Miyazaki are obliged to give rather long exposures, without loss of resolution, which proves they benefit from particularly stable images. Their prints are very rich in subtle half-tones but, in my opinion, those of Miyazaki, superb to look at, lose a great deal through reproduction in the various magazines (as in *Sky and Telescope*). On the other hand, it is certain that the Newtonian telescopes of the two best planetary photographers provide more contrasty images than those which could be obtained with Cassegrains or Schmidt–Cassegrains. I conclude that if the development technique of Parker and Miyazaki suits them, it need not suit others. At Lowell Observatory, Flagstaff, I used Rodinal developer at 1:50 dilution (14 min at 20 °C) in comparison with HC-110 (dilution A, 6–7 min at 20 °C) on hundreds of Mars and Jupiter negatives. For Jupiter, I found that the contrast obtained with Rodinal was inadequate and the level of fogging was obtrusive. Today I am using HC-110 (dilution B, 12–14 min) as the preferred developer for Mars and Saturn, and perhaps D19 or Dektol for Jupiter (more often than not with a 'small' 200–250 mm telescope with a large central obstruction).

At Cotonou (1980–4) I obtained good results with D19 (5–6 min at 20 °C), but this took less advantage of the sensitivity of TP 2415 than did HC-110 (dilution B, 14 min at 20 °C). For further details, see pp. 49–53 and Tables 3.9 and 3.12.

When the exposure time has to exceed 2 s, it can be useful to hypersensitise the film with forming gas (40–46 h at 3 psi at 55 °C).* This hypering has several advantages: (1) it allows us to halve the exposure time (though only for exposures of 2–8 s or longer); (2) the contrast of TP 2415 is slightly lowered, and the toe of the sensitometric (or characteristic) curve shows an improvement in the recording of shadow detail. This is advantageous, as after hypersensitisation by forming gas one must develop the film in D19, this being the only developer which does not increase the level of fog.

TP 2415 film, which is described as panchromatic, has a very individual spectral sensitivity: very sensitive in the red, of low sensitivity in the green and blue, a little more sensitive in the violet. The use of coloured filters, so often very important in planetary photography is therefore a rather delicate matter: red filters (W25, 29) lead to just a small increase in exposure times; on the other hand, blue filters (W47, W47B) entail much longer exposures (so much so as often to be incompatible with the goal of high resolution!). Therefore, it is best to use violet filters in place of blue filters. For example, if the exposure time with no filter is 2 s, it will be 3–4 s with a red filter and 50 (!) s with a W47 blue filter (for further details, see p. 41). Miyazaki therefore uses a violet B340 filter (comparable with the Wratten 36).† With a 'small' 200–300 mm telescope the use of blue filters remains questionable. At Flagstaff or at the Pic du Midi, thanks to the use of low focal ratios ($F/D = 20$–32), I was able to get acceptable results with the Wratten 47 filter, and the exposure times were only 2–6 s. For an aperture of 200–250 mm the red filter may be replaced with an orange one for both Jupiter and Mars (the exposure times remain acceptably low).

When one has access to a telescope at least

---

* I have always thought that 'hypersensitisation' is an inappropriate term. For 'normal' photographic exposures (1/1000–1 s), the speed of the film is unchanged by hypering (or 'baking'). The value of hypering is in reducing the reciprocity failure for longer exposures. It can also modify the contrast. Note that the stated gas pressures (such as '3 psi') are taken to imply additional pressure, over and above normal atmospheric pressure (1 atm = 15 psi).

† Another reason Miyazaki and Parker have tried this B340 filter for Mars is that the W47 filter has what they refer to as a 'red leak', which in view of the high sensitivity of TP 2415 to red light could cause anomalous 'Blue Clearings' to be recorded by photography through the W47–TP 2415 combination. I understand that the B340 filter does not have this same problem. (RJM)

400 mm in diameter, and wishes to obtain scientifically worthwhile documents, one must use at least two colour filters, a red and a blue, as they can yield important scientific information as much for Mars as for Jupiter. The red one must be either a W25 or a W29, and the blue one should be a W47 or a W47B. The blue may sometimes be replaced by a violet one (B340 or W36). The use of filters is much easier with a CCD camera, as their sensitivity remains almost constant across the spectrum from the UV to the IR.

In planetary photography it can be very important to be able to time each negative to a precision of 5–10 s for Jupiter and perhaps 30 s for Mars. Today most reflex cameras of the 24 × 36 mm format can be equipped with daters which imprint the date and time (hours and minutes, but sadly not seconds) upon the film. The main job being done by the dater, it only remains for the photographer to mark each strip of 4–6 negatives after development with the year, month and filter used. For example, the negative might bear the figures 2 03 45, i.e. the second day of the month, 03h 45m UT. It would then suffice to add, with a fine indelible pen, 1992 October, W25 on each strip of 4–6 negatives. When using an old-fashioned camera (an old Olympus model, or Leica M, Robot, Miranda, Laborec, Exakta, etc.), one has to note the details of the times on a dictaphone (which can be worn round the neck on a strap) or a tape-recorder (or by hand, of course). It is a tedious and delicate operation, as one must not make a mistake!

Once developed and dried, the negatives should be examined at leisure with a magnifying glass. A ×10 lens suffices for planetary negatives (for focal ratios F/D above 120, a ×5 lens would be better). This examination allows the most promising negatives to be marked in a convenient manner.

In the past, one automatically composited all planetary photographs, at Flagstaff as well as at Catalina or Pic du Midi. Between 4 and 12 images, of as nearly comparable quality as possible, were printed by superposition on a fine-grain film (see p. 57). Compositing brings a gain in resolution and contrast, absolutely vital for the grainy, low contrast films of that time.

After TP 2415 film came into general use, hardly any compositing was done. The quality of the prints obtained could not be improved upon further. It is perhaps an injustice to renounce compositing, as a result of the improved quality of emulsions today. The technique is returning to favour to some extent, being much easier with video images.

The printing of planetary negatives is much easier than that of lunar negatives – above all, with those of Mars, which require little harmonisation. Over-enlargement must be avoided at all costs. Failure to recognise this has led to the downfall of many beginners. According to the resolution of the negative, it should be enlarged to a diameter between 30 and 60 mm (in exceptional cases 80–100 mm). A compromise size must be found which is not large to reveal the grain, yet which is large enough to show all the details on the negative easily. One can either opt for a constant enlargement factor (×10 or ×12, for example), the factor varying according to the angular diameter of the planet, or for a constant diameter (and thus a variable degree of enlargement) when one wishes to facilitate later measurements from the prints. Mars can be readily printed on Ilford Multigrade III, on grades 3–5 (according to the processing of the negative). A disk of carboard placed at the end of a fine rod allows the print to be 'dodged' or harmonised by overexposing the darker markings, when trying to bring out details in the lighter, desert areas of the planet. An unsharp mask can play a useful part in harmonising the print, but remains a delicate tool (see p. 57 and Figs. 4.38 and 4.44).

For Jupiter, the negative having a larger diameter, the enlargement factor can be less. The polar diameter of the planet can be varied between 40 mm and 80 mm (50–60 mm on average). It is essential to harmonise the prints, as the limb darkening of the planet and the falling-off of light on the phased side will be found to lead to the print being too dark at its borders (especially when photographing the planet in red light). The edge of the planet therefore needs to be masked for most of the exposure during printing, with the help of an opaque disk pierced with a small elliptical hole which is placed in the light-path of the enlarger which masks the edges of Jupiter's disk; the mask is moved continuously during the exposure. This type of harmonisation is hard to achieve, as the instant for which the limb is exposed necessarily varies from one negative to the next. A negative taken in blue light, a little overexposed, requires very little harmonisation (the mask need be present for only about 20% of the total exposure time), whereas an underexposed red-light negative can be very difficult to harmonise, requiring masking for up to 70% of the total exposure time (which makes it impossible to obtain a completely black sky background). Each image must be judged on its own merits. For a normal negative exposed without a filter, the mask needs to be present for about 50% of the total exposure time (assuming that the centre of the image needs 40 s, the mask must be introduced after 20 s, in order to cut off the outer 20% or so of the equatorial diameter; it must be kept in continual motion in all directions). At the same time, one can correct for the phase effect, by preferentially underexposing the phased limb. D. Parker and especially I. Miyazaki have tended to make their prints on relatively 'soft' printing paper, which presents a great richness of subtle half-tones. They are superb to look at but give, most often, poor results after publication. The French, English and German photographers produce contrasty prints which lend themselves more to magazine publication. I myself use grades 4 and 5.

4.4.4 Planetary photography in colour

As I have already stated several times, I am not keen on high resolution colour photography: it involves a loss in both contrast and definition, complex processing and expensive publication in scientific journals. In any event, the clouds of the planet Venus are monochromatic (to be photographed only in UV light) and Mars shows no more than one clear tone: the ochre of the dusty deserts. The blue-green of the darker regions is mostly a complex grey which is, anyway, poorly captured by colour photography. The only advantage is that often the white clouds are more clearly recorded on colour photographs than on black and white unfiltered photographs, as D. Parker has ably demonstrated.

On the other hand, colour photography of Jupiter can be very interesting indeed. G. Kuiper and his team (Kuiper, 1972; Larson *et al.*, 1973; Fountain & Larson, 1971) thoroughly studied the possibilities of colour photography, with the 1 m.50 reflector at Catalina. After numerous trials, the American astronomers used Ektachrome (from 1968–9) which was deliberately underexposed at 25% of the normal time (i.e. for 1/8 s on Jupiter at $F/D = 75$). The film was overdeveloped in a controlled fashion. This led to an increase in grain as well as a slight drift in the colour balance. The negatives, being underexposed, were very dense and saturated. They had to be rephotographed in order to obtain a normal density and to correct the colour balance. Kuiper and his co-workers obtained some 12 000 photographs of Jupiter in colour, of which 500 were selected to be printed onto large-format colour slide film or onto paper. The resolution of

these colour photographs was very good, as can be seen from the plates published in the *Communications* of the Lunar and Planetary Laboratory. On many of the colour slides the resolving power reached nearly 0″.20. An important point: the sharpness of the image is not constant over the whole of Jupiter. That is why several slides must be studied, taken over short intervals of time, in order to get an exact idea of the various Jovian clouds. Kuiper asserted, with good reason, that the colour gave important information about the chemical composition of the various cloud formations of Jupiter and Saturn.

Today, I think that even better results could be obtained by the trichrome method, thus benefiting from the resolving power and the contrast of TP 2415 film (used with red, blue and green filters). The reader will certainly have seen the superb Jupiter images obtained by the Hubble Space Telescope, by the trichrome method in conjunction with a CCD camera. D. Parker has obtained some very beautiful colour images of Jupiter with his CCD camera.

The amateur equipped with a 300–400 mm telescope could try out the new professional colour films (Ektachrome, Fujichrome), underexpose them systematically, overdevelop them and recopy the diapositive, made deliberately too dense, on colour films specially designed for duplication (in order to recapture the normal density by overexposing the copies).

4.4.5 Special points about individual planets

Venus

Photography of the planet Venus is not too difficult if one is content to record only the phases, as, in view of its great brightness, Venus can withstand considerable enlargement (provided that the image quality permits). The exposure time, for a focal ratio of $F/D = 100$, reaches 1/30 s on TP 2415 film, which is excellent if one has the use of a vibration-free Compur-type shutter. On the other hand, if one has to make the exposure manually by means of an external shutter (see p. 39), this exposure time is clearly inappropriate. One must therefore use a very dense blue or green filter in order to increase the exposure time to 0.5 s or 1 s, which is more realistic for this type of exposure. Unfortunately, the turbulence rarely enables Venus photographs with really sharp cusps to be obtained (except at Cotonou, Fig. 4.34). It is generally preferable to choose those times when Venus is high up in the sky, in preference to the

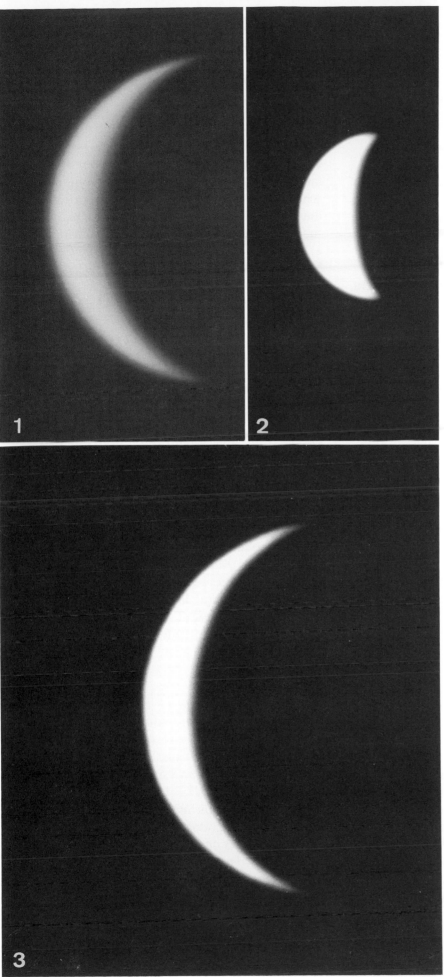

Figure 4.34. The planet Venus. *1. 1967 May 6 at 21.30 UT by G. Nemec with his 200 mm OG. The sharpness of the cusps is evidence of the good seeing in Munich. 2. 1981 October 6 at 17.45 UT, by the author at Cotonou with the C14 (355 mm Schmidt–Cassegrain); F/D = 60; on TP 2415 film. 3. 1981 December 10 at 19.01 UT, by the author, with the C14, in very good seeing (note the extremely fine cusps).*

morning (when there is better thermal equilibrium between the telescope and its surroundings). However, this type of photograph is of little interest.

Photography of the cloud formations in the atmosphere of Venus can be attempted with the help of a UV filter. It was, in effect, with the aid of a 26 cm Newtonian in 1957 that C. Boyer discovered the four-day retrograde rotation of the atmosphere of Venus. This important discovery could be made thanks to the high quality of the images at Brazzaville (Congo) and the use of a strong blue filter. Later, Boyer and his co-workers (Camichel and Guérin, among others) used, with success, the Schott UG5 glass filter. Significant results can be obtained with a Wratten 36 filter (either on its own or in conjunction with a Wratten 18A). Focusing can never be done with the filter in place. A less absorbent filter (clear yellow) must be available, either of gelatine or glass, of the same thickness as the UV filter: focusing is carried out with the transparent filter in place; then the UV filter is substituted by appropriate means and the photographs taken through the UV filter.

With photographic apparatus of the type described by D. Parker, or that designed by myself (see p. 36), it is better to place the UV filter just in front of the film, inside the camera, and place the transparent filter a little in front of the focusing screen, in order that the displacement of the image due to the passage through the filter be identical for the two optical paths. In this way Venus can be viewed continuously, which enables one to expose at the moments when seeing is best.

For the range of phases around dichotomy (the most interesting), one can work at $F/D = 70-80$ on TP 2415 film. The exposure time will vary from 1/10 s to 2 s according to the type of filter used, the W47 being less absorbent than the Schott UG5. The atmospheric formations being of low contrast, the TP 2415 film should be developed in a highly energetic developer, to infinite γ (D19, MWP2 or Dektol). If the TP 2415 film exhibits too low a sensitivity in the violet, one can try an orthochromatic film hypersensitised in forming gas, such as Kodak's Ektographic or Agfaortho 25. There is much of real scientific value in photographing Venus in UV light in this way, although it is an operation which attracts few amateurs.

Mars

Mars can be photographed only when it makes a close approach to the Earth, i.e. for periods of 3 or 4 months, once every 2 years. In reality the situation is worse than this, since only the perihelic oppositions are really fruitful. See Figs. 4.35–4.39.

As can be seen from Table 4.3 (p. 64), Mars will not be favourably observable from our latitudes (40–50° N) until the beginning of 1995, when it will appear high in the sky but its diameter will be very diminutive. The planet will make a close approach to the Earth during 2003.

In view of its small apparent diameter (16–25″), Mars should be photographed at focal ratios $F/D = 80-150$. Its albedo, relatively high, allows us to use exposure times between 1 s and 4 s (with a red filter). The use of a blue (W47) filter is very interesting with Mars, as it allows the polar hoods, certain white clouds and the mysterious 'Blue Clearing' to be recorded (when visible). Unfortunately, TP 2415 film is less sensitive to blue light and requires excessive exposure times with a W47 filter. The blue images not necessitating a very high resolution, one can reduce the focal ratio to 40–50 and thus expose for only 6–7 s, developing in HC-110, dilution B 14 min, or in Dektol for 3 min. A longer exposure requires good images, an excellent drive and perfect polar alignment. That is why D. Parker prefers to photograph Mars in colour (on Ektachrome 100 or 200) and then to rephotograph the negative thus obtained, with the help of a W47B blue filter. By this means one obtains the same result as direct photography in blue light. Actually, I think that it is likely that

Figure 4.35. The planet Mars, *photographed at average resolution. 1. 1971 August 16; 2. 1971 August 13; 3. 1971 August 9; all by Nemec, with his 200 mm OG. 4. 1988 September 21, by G. Thérin, with a C8; F/D = 100; 1 s. 5. 1988 October 1 at 07.35 UT, by the author with the great refractor of the Lowell Observatory (diaphragmed to 450 mm); F/D = 44; 0.25 s on TP 2415 film. 6. 1988 September 16, by the author, with the same refractor (the seeing still being mediocre). 7. 1990 December 6 at 19.38 UT, by B. Flach-Wilken (Wirges) with his 300 mm Schiefspiegler; F/D = 170; 4 s on TP 2415 film. 8. 1988 October 21 at 21.22 UT, by B. Flach-Wilken with his 300 mm telescope; F/D = 170; 4 s on TP 2415 film. 9. 1990 October 14 at 03.21 UT, by Wolf Bickel, with a 404 mm Newtonian, CCD camera; 0.32 s (very good seeing). The small arrows indicate several details of interest.*

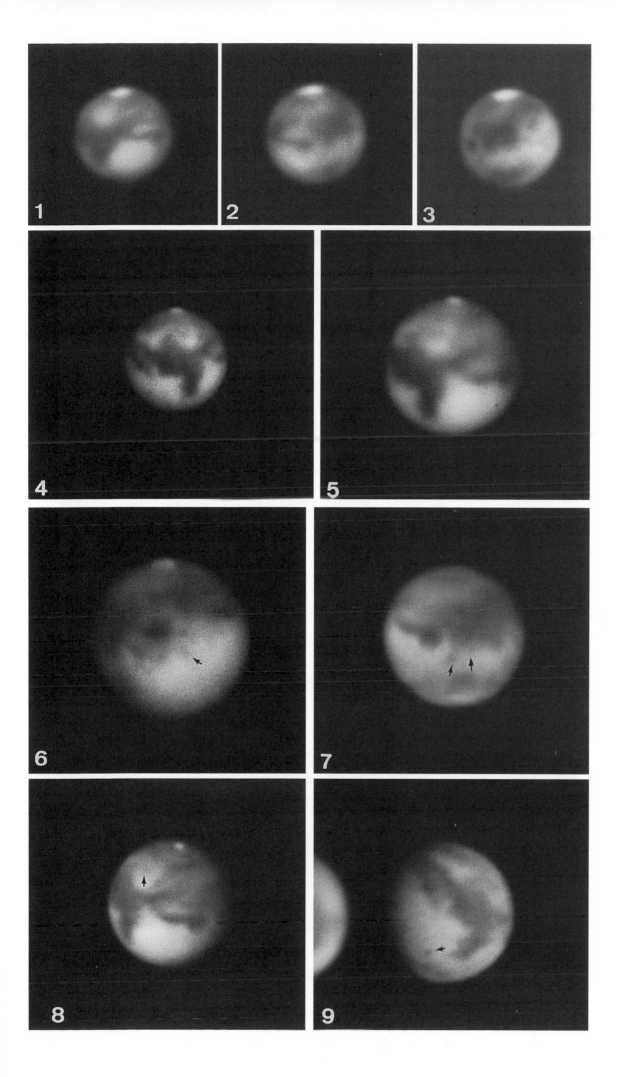

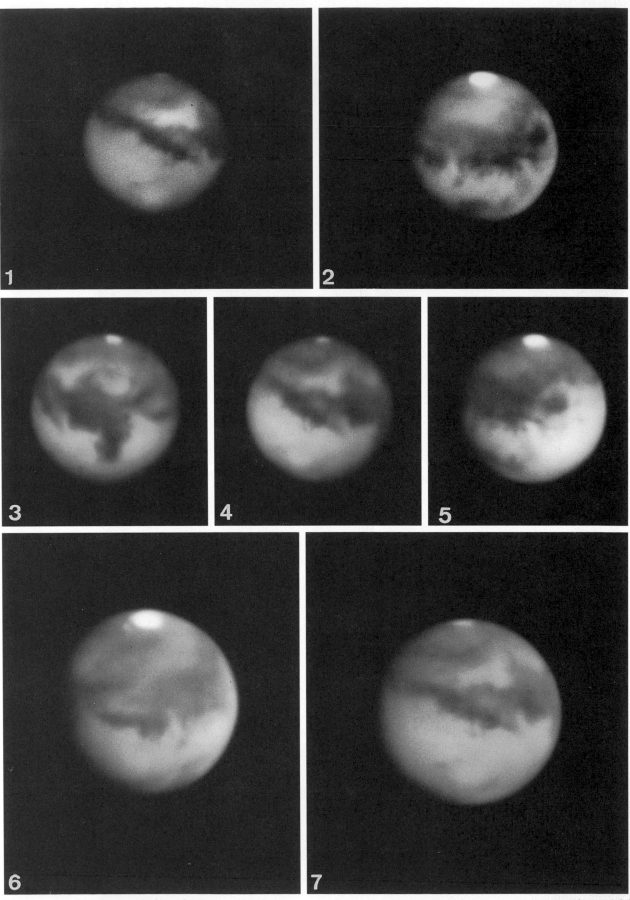

Figure 4.36. Photographs of Mars at high resolution. *1. 1988 September 27 at 22.51 UT (ω = 219°), by G. Viscardy, with his 520 mm Cassegrain; F/D = 70; 1 s on TP 2415 film. 2. 1988 September 11 at 00.44 UT (ω = 29°), by G. Viscardy (same data as before). 3. 1988 September 30 at 05.10 UT (ω = 286°), by D. Parker, with his 400 mm Newtonian; F/D = 165; 2 s on TP 2415* *film. 4. 1988 October 4 at 03.59 UT (ω = 233°), by D. Parker, under the same conditions. 5. 1988 September 18 at 07.16 UT, also by D. Parker (ω = 62°), same technical data as before. 6. 1988 September 4 at 18.30 UT (ω = 352°), by Isao Miyazaki, with his 400 mm Newtonian; F/D = 175; 2 s on TP 2415 film. 7. 1988 September 16 (ω = 219°), by I. Miyazaki, same conditions as before.*

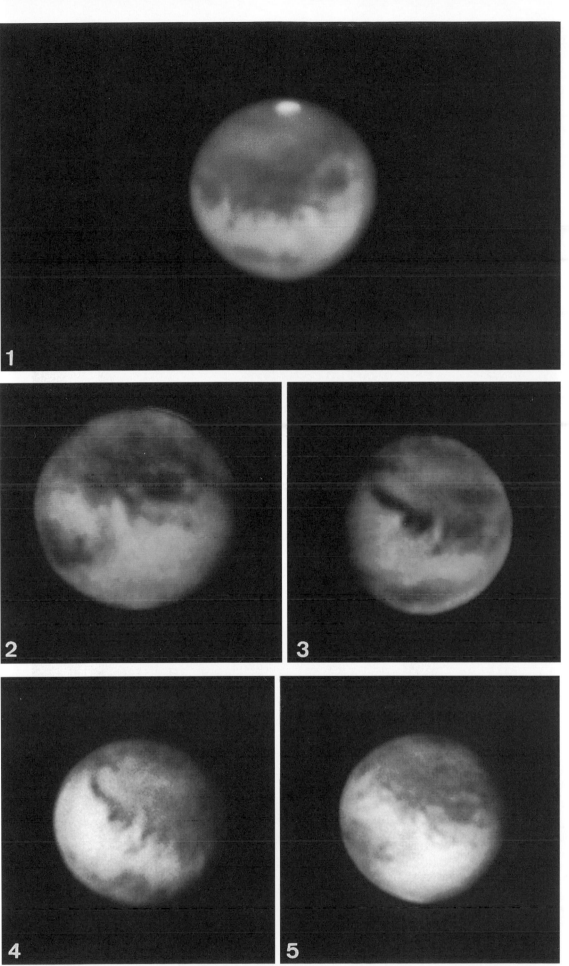

Figure 4.37. The planet Mars at high resolution.
1. 1988 October 2 at 14.47 UT ($\omega = 49°$), by I.
Miyazaki, with his 400 mm Newtonian; F/D = 175;
3 s on TP 2415 film. 2. 1990 December 5 at 23.08
UT ($\omega = 88°$), by Terry Platt, with his 320 mm
telescope and CCD camera. 3. 1990 November 6 at
00.15 UT ($\omega = 0°$), again by Terry Platt and his
CCD camera. 4. 1990 December 6 at 19.05 UT ($\omega = 11°$), by Wolf Bickel, with his 404 mm Newtonian
and CCD camera; 0.5 s with image processing. 5.
1990 December 6 at 22.49 UT ($\omega = 66°$), by W.
Bickel, same conditions as before.

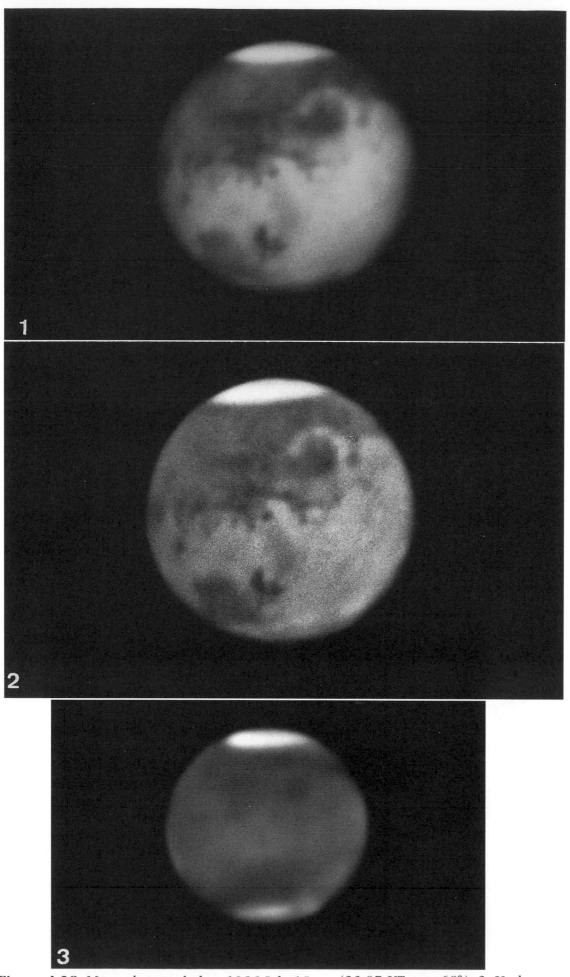

Figure 4.38. Mars, *photographed on 1986 July 15 with the 106 cm Cassegrain telescope at the Pic du Midi Observatory, by the author. Excellent images, F/D = 50; 0.50 s on TP 2415 film. The image of Mars measured no more than 5 mm across on the negative. 1. Wratten 29 red filter and normal printing* (23.07 UT; ω = 68°). *2. Unsharp-masked print from the same negative. 3. Wratten 49 blue filter; F/D = 32; 4 s on TP 2415 film (23.09 UT; ω = 68°). A slight 'Blue Clearing' allows us to see the outlines of the albedo features. Note the hood over the N pole and the cloud around the limbs.*

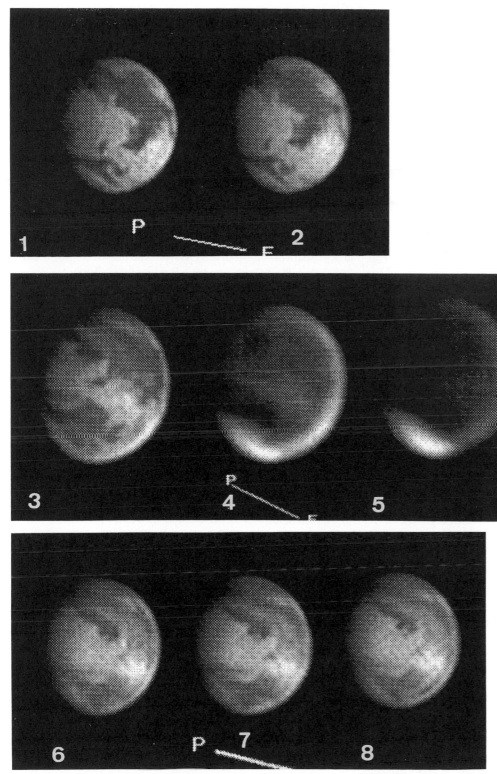

Figure 4.39. Mars in 1992 imaged by D. Parker with his CCD camera on his 400 mm Newtonian at Coral Gables, Florida. 1, 2. 1992 November 7 at 08.54 and 09.26 UT, with Wratten 25 red filter ($\omega = 259°$ and $267°$); F/D = 48; 0.20 s. 3–5. 1992 October 25 with W25 red filter at 08.11 UT (3), then with W58 green filter (4) and finally with W47 blue filter (5) ($\omega = 11–18°$). 6–8. 1992 October 30, 09.52–10.10 UT, with W25 red filter ($\omega = 349–352°$); F/D = 48; 0.16 s. P = preceding; F = following. These images show the extraordinary possibilities of the CCD camera, the diameter of the planet being no greater than 9″.9 at the time the pictures were taken.

Ektachrome is more sensitive to blue light than is TP 2415, but the contrast and resolving power of colour films remain low. Therefore it would be better to photograph Mars in blue-violet light on Type 103aO film, very sensitive to blue light and very contrasty, using a high focal ratio ($F/D = 180$–220) to avoid loss of resolution due to the grain of the film in question. One can again try using hypered blue-sensitive films, such as Ektagraphic (Kodak) and Agfaortho 25.

The details of the Martian surface are very contrasty, from which it follows that one is not obliged to develop TP 2415 film to an infinite γ rating. It will be preferable to use HC-110 (dilution B) for 10–12 min (at 20 °C) or Rodinal 1 : 50 (or 1 : 100) for 12 min at 20 °C. Printing can be carried out on multigrade paper (grades 3–4) with slight harmonisation (shading the darker markings to record detail in the light regions). Each print should be accompanied by the data needed to make scientific use of the photograph: date in UT (year, month, day, hour, minute), omega (longitude of the Central Meridian), apparent diameter, latitude of the centre of the disk, phase. All these data can be found in the B.A.A. Handbook.

Really good Mars photographs are of much scientific interest. Those obtained by G. Viscardy, D. Parker, I. Miyazaki and T. Platt are equal in standard to the photographs obtained in the specialised observatories, and often better than those not so long ago obtained at the NASA stations or at Lowell Observatory. It is therefore important that the best of such documents should be carefully preserved, for example in the Planetary Documentation Center at Flagstaff. It should not be forgotten that Dr R. J. McKim of the B.A.A. and Dr D. Parker of the A.L.P.O. will always be happy to receive useful Mars photographs (or other observations), to enable them to publish informative reports about the changing face of the planet.*

Jupiter
Jupiter represents the favourite high resolution target for the amateur, for three equally valid reasons: it has a large apparent diameter, it has

frequent oppositions (once per year) and its atmosphere is constantly changing. Unfortunately, Jupiter also spends some periods in negative declination, at the time of its summer oppositions (1999–2003).

Jupiter exhibits an apparent diameter at least twice that of Mars (an average of 47″ against 19″). It is therefore easier to photograph with 200–250 mm telescopes. Its albedo is always lower than that of Mars, obliging us to use lower focal ratios ($F/D = 60$–90, for exposures between 1 and 3 seconds, or a little more with a red filter). The use of a blue filter is of less interest with Jupiter, unless one wishes to obtain precise positional measurements of the various spots: the Great Red Spot appears better defined at short wavelengths, and the borders of the disk will appear sharper. With the Wratten 47 filter, the focal ratio should be reduced to about 40 and an exposure of 8–10 s should be given: these are difficult conditions, unless the images are perfect and the drive excellent.

The details of Jupiter's atmosphere often show low contrast (especially in red light); it is therefore necessary to develop the negatives energetically. I would suggest either HC-100 (dilution B) for 12–14 min at 20 °C or D19 or D19B for 5–6 min at 20 °C. My recommendations oppose the techniques employed by Parker and Miyazaki but I think that the final choice in development technique depends upon the instrument used. It is evident that the Newtonian, with its small central obstruction will furnish much more contrasty images than the more complex Schmidt–Cassegrain, with a large central obstruction. The type of local turbulence also comes into play. It is for each to find the technique that best suits him (for more details, see p. 49). See Figs. 4.40–4.45.

Jupiter exhibits changeable cloud formations and turns rapidly about its axis; it is thus very important that each photograph be accompanied by precise numerical data on the year, month, day, hour and minute. Following from the rapid rotation rate of Jupiter, one will not be able to obtain precise positional measurements if one has not calculated the longitude of the central meridians ω_1 and ω_2 to an accuracy of 1 min or better (which is within the capacity of film daters). The absence of seconds on film daters is a drawback, but not a serious one, as the error will not exceed 59 s and will be only 30 s on average. For making prints of Jupiter negatives, see p. 115.

The planet being essentially in a constant state of change, each high resolution photo (showing details smaller than 1″) can constitute a precious

---

* The present addresses of these Mars recorders are as follows:
  Dr. Richard J. McKim (Director of the BAA Mars Section)
  5 Ashton Road, Oundle, Peterborough PE8 4BY, Great Britain.
  Dr Donald C. Parker (senior Mars recorder of the ALPO)
  12911 Lerida Street, Coral Gables, Miami, Florida 33156, U.S.A.

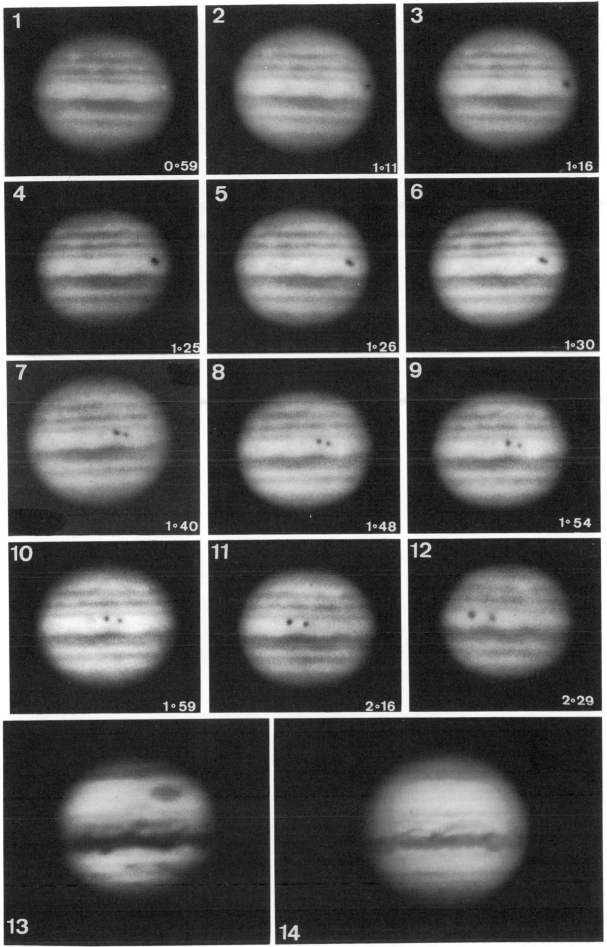

Figure 4.40. The planet Jupiter. *1–12. Historical document showing the simultaneous transit of the satellites Io and Europa and their shadows, 1967 January 31, from 00.59 to 02.29 UT, by G. Nemec, with his 200 mm refractor; F/D = 130; 6 s on Addox KB 14 film. The shadow of Europa precedes the shadow of Io. 13. 1990 January 9 at 20.14 UT, by F. Kufer, with a 300 mm reflector; F/D = 117; 2 s on TP 2415 film. ($\omega_1 = 324°$; $\omega_2 = 21°$) 14. 1989 November 20, by S. Deconihout, with a 320 mm Cassegrain; F/D = 90; 3 s on TP 2415 film.*

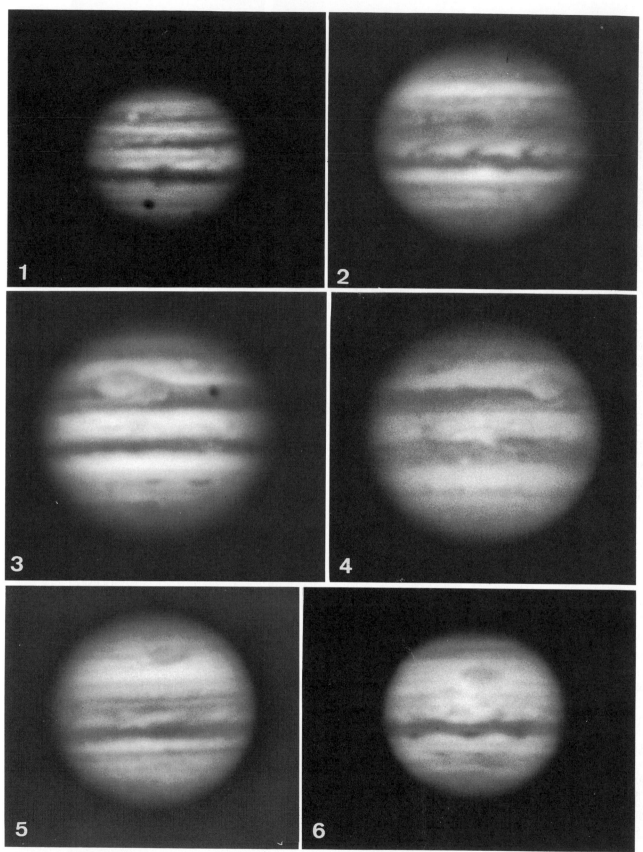

Figure 4.41. The planet Jupiter, *viewed by amateurs. 1. 1981 March 16 at 01.12 UT, by the author, with his C14 (355 mm); F/D = 70; 2 s on TP 2415 film (ω₁ = 339°; ω₂ = 139°) 2. 1991 February 19 at 21.40 UT by B. Flach-Wilken, with his 300 mm Schiefspiegler; 30 m focal length (ω₁ = 11°; ω₂ = 210°) 3. Undated image by B. Flach-Wilken. 4. 1988 October 29 at 22.32 UT, by* B. Flach-Wilken, with the same telescope, 34 m focal length; 1.5 s on TP 2415 film (ω₁ = 196°; ω₂ = 347°) 5. Undated photograph by C Arsidi, with his 225 mm Takahashi Schmidt–Cassegrain; F/D = 120; on TP 2415 film. 6. 1990 January 18 at 22.44 UT, by G. Thérin, with a C8 (203 mm); F/D = 100; 3 s on TP 2415 film (with yellow filter) (ω₁ = 25°; ω₂ = 13°)

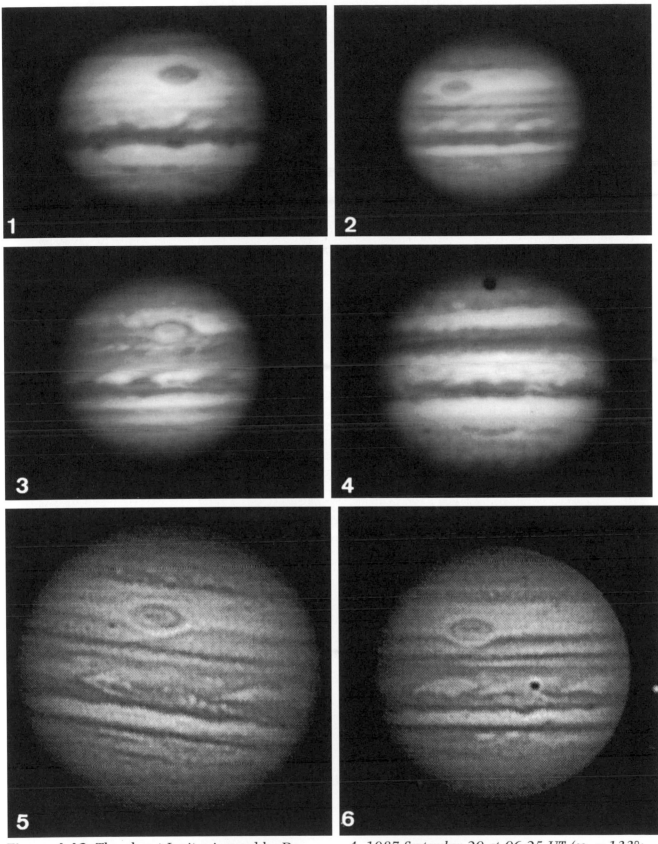

Figure 4.42. The planet Jupiter imaged by D. Parker. *1. 1990 January 6 at 02.49 UT ($\omega_1 = 293°$; $\omega_2 = 19°$); 400 mm reflector; F/D = 93; 1 s on TP 21415 film. 2. 1992 May 14 at 14.35 UT ($\omega_1 = 303°$; $\omega_2 = 56°$), with the same telescope; F/D = 144; 4 s on TP 2415 film. 3. 1990 December 31 at 05.51 UT ($\omega_1 = 171°$; $\omega_2 = 37°$); same data.*

4. 1987 September 20 at 06.25 UT ($\omega_1 = 133°$; $\omega_2 = 140°$), with a 310 mm reflector; F/D = 123; 2.5 s. 5. Undated CCD image. 6. 1992 March 2 at 04.22 UT ($\omega_1 = 193°$; $\omega_2 = 41°$); 400 mm reflector; CCD camera; white light; F/D = 27; 0.47 s. Admirable resolution.

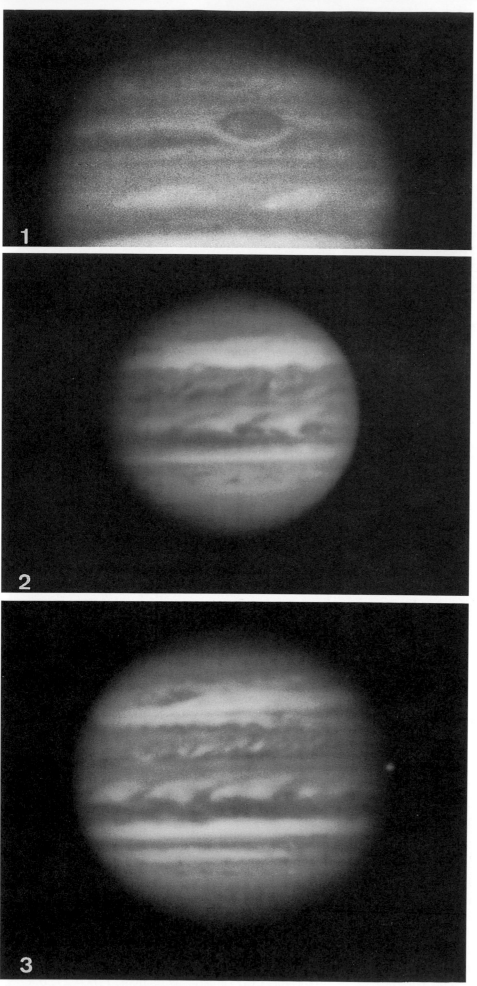

Figure 4.43. Jupiter photographed at the most favourable sites. *1. 1976 November 17 by L. Tomas, with the 400 mm solar telescope of the Observatoire del Teide (Tenerife), on the film of that epoch. 2. 1990 November 18 at 19.30 UT ($\omega_1 = 357°$; $\omega_2 = 187°$), by I. Miyazaki, with his 400 mm Newtonian; F/D = 144. 3. 1991 March 7 at 12.25 UT ($\omega_1 = 40°$; $\omega_2 = 120°$), by I. Miyazaki, with his 400 mm; F/D = 144; 3.5 s on TP 2415 film. The last image shows an excellent resolution of better than 0″.30.*

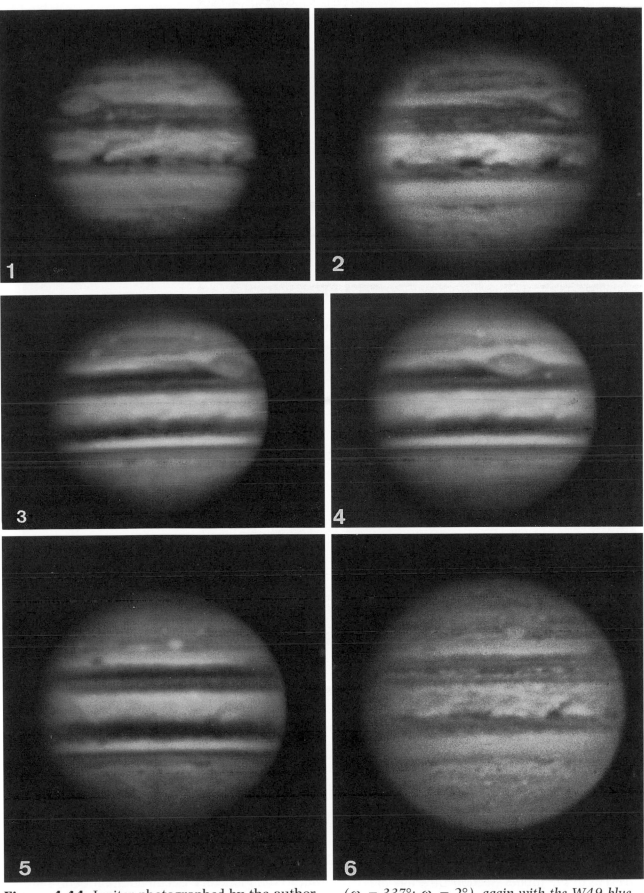

Figure 4.44. Jupiter photographed by the author at the Pic du Midi Observatory. *1. 1986 July 21 at 01.13 UT ($\omega_1 = 32°$; $\omega_2 = 50°$), with the 106 cm Cassegrain; $F/D = 32$; W29 red filter; 1 s on TP 2415. 2. 1986 July 20 at 03.14 UT ($\omega_1 = 307°$; $\omega_2 = 333°$); W29 red filter; 1 s on TP 2415 film. 3. Several seconds after No. 2, with W49 blue filter and 8 s on TP 2415 film. 4. 1986 July 20 at 04.02 UT* ($\omega_1 = 337°$; $\omega_2 = 2°$), *again with the W49 blue filter. 5. 1986 July 10 at 03.47 UT ($\omega_1 = 188°$; $\omega_2 = 279°$), with W49 blue filter. 6. Several seconds before No. 5, with W29 red filter and 1 s exposure, an unsharp-masked print made by Tomio Akutsu. (The result is striking: the unsharp mask improves the contrast ratios and the resolution seems higher.)*

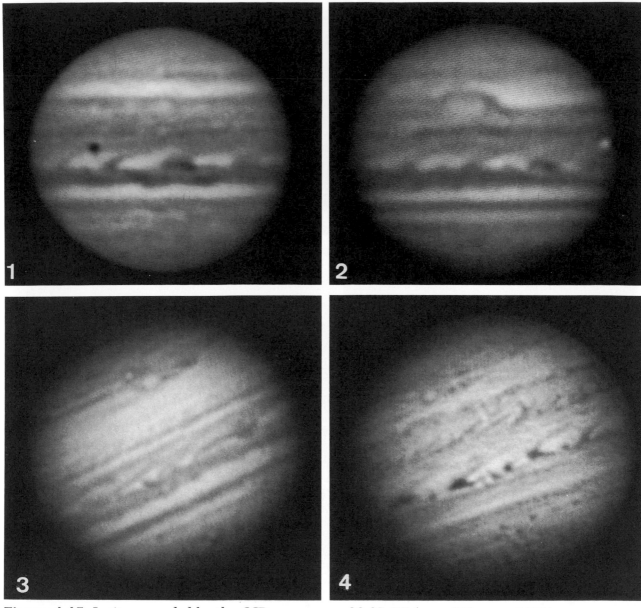

Figure 4.45. Jupiter recorded by the CCD camera. *1. 1991 February 26 at 22.51 UT ($\omega_1 = 80°$; $\omega_2 = 226°$), by Terry Platt, with his 320 mm quad-Schiefspiegler. 2. 1991 February 1 at 22.12 UT ($\omega_1 = 67°$; $\omega_2 = 43°$); same photographer and conditions. 3. 1990 December 6 at 03.00 UT ($\omega_1 = 76°$; $\omega_2 = 134°$), by W. Bickel, with his 404 mm and CCD camera; RG 630 (red) filter; 1.5 s exposure followed by image processing. 4. 1992 March 4 at* 22.27 UT ($\omega_1 = 91°$; $\omega_2 = 278°$); same photographer and equipment, but a composite of 22×0.08 s exposures, and with compensation for the planet's rotation. These two last images exhibit an astonishing resolution, especially that for March 4. At Bergish-Gladbach, and with a telescope of only 404 mm aperture, the German amateur obtains images as good as those recorded, only a few years earlier, from Pic du Midi and Catalina Observatories.

document. The best photographs of Jupiter should be entrusted to the Jupiter Section Directors or Recorders of national organisations: Dr J. H. Rogers in Great Britain and P. W. Budine or J. Olivarez in the USA.*

Documents of professional quality should be sent to the Planetary Documentation Centre at Flagstaff, to be archived (Lowell Observatory,

* Current addresses of these individuals are as follows:

Dr John H. Rogers (Director of the BAA Jupiter Section), 10 The Woodlands, Linton, Cambridge CB1 6UF, England

Philip W. Budine (ALPO Jupiter recorder), R.D.3, Box 145C, Walton, New York 13856, USA

José Olivarez (ALPO Jupiter recorder), 1469 Valleyview Court, Wichita, Kansas 67212, USA

1400 West Mars Hill Road, Flagstaff, Arizona 86001, USA). For photographing Jupiter in colour, see p. 116.

Saturn

The most beautiful of all the planets, distant Saturn (Fig. 4.46), is hard to photograph well, since it is small and faint. Owing to its distance from the Sun, Saturn requires an exposure time three times longer than for Jupiter, together with a very high focal ratio. At $F/D = 100$ Saturn demands an exposure time of 4–5 s on TP 2415 film; for focal ratios of $F/D = 160$–200, one can reach an exposure time as long as 20 s, but such times are always to the detriment of resolution. To photograph the Cassini Division, which separates ring A from ring B, with certainty, in a 5 s exposure, one needs both excellent seeing and perfect drive (as the image stability must be at least $0''.5$ in 5 s of time). In effect, the basic test for evaluating a photograph of Saturn consists in seeing whether or not the Cassini Division is well defined (which is, however, observable visually with a 60 mm refractor). It is surprising that it should be so easy to photograph a lunar crevice $0''.5$ in width when one hardly ever obtains a sharp image of Cassini's Division, even when Saturn's rings are exhibited at their most favourable tilt. It is a case of exposure times: having overcome the turbulence and getting the telescope drive to work perfectly for a period of 1 s, it is very difficult to achieve the same result during an exposure five times as long. Saturn presents another difficulty: the global details (dark belts, bright zones, polar shadings) require high contrast for them to be correctly recorded, whereas a more modest contrast is required for the ring details, and so a balance must be struck. I think that such a compromise is needed, for Saturn pushes to the maximum the sensitivity of TP 2415 film. The ideal would therefore be to hypersensitise TP 2415 in forming gas, a treatment which slightly reduces the contrast and allows the exposure to be cut by half (2.5 s instead of 5 s, etc.). The film will then be developed to the maximum, in HC-110 (dilution B) for 12 min at 20 °C (or even 14 min if the negative is too light).

Printing photographs of Saturn poses no particular difficulty. A slight harmonisation allows one to reduce the difference in density between the globe and rings and between the equatorial and polar regions of the planet.

It is almost impossible for the amateur to obtain a valuable result with Saturn, unless he or she happens to record the unexpected appearance of a Great White Spot in the equatorial or tropical regions; this actually happened in 1990. Saturn is best seen from our latitudes from late autumn through spring, as it then has a positive declination (as from 1997 to 2003).

Mercury, Uranus, Neptune and Pluto

These four planets are of little interest to high resolution astrophotography, unless we content ourselves with recording the position of the principal satellites of Uranus and Neptune or the position of Pluto wandering among the stars. This is not, properly speaking, high resolution as it is necessary to work at the prime focus with at least a 25 cm telescope. It is preferable to use TP 2415 film, hypered with forming gas, as the exposure times can be long.

The five main satellites of Uranus (Ariel, Umbriel, Titania, Oberon and Miranda) exhibit stellar magnitudes ranging from 14 to 16.5. Titania and Oberon can be photographed very easily, being far from the planet. The exposure times vary considerably, according to the aperture used, the focal ratio ($F/D = 5$–15 or more) and the sky transparency. One can try exposure times of 1–4 min (with TP 2415 film developed for 6 min in D19 at 20 °C). The other satellites are more difficult to photograph, the more so as an exposure of 4–10 min runs the risk of greatly overexposing the image of Uranus. However, amateurs very regularly photograph four of the satellites of Uranus. Owing to the length of exposure, it is absolutely necessary to correct for drift of the field by using a high power guiding eyepiece (preferably an off-axis guiding eyepiece) or, even better, with the help of a beam-splitter system, allowing the exposure to proceed while the image is still being scrutinised against the eyepiece graticule.

The two major satellites of Neptune (Triton and Nereid) have magnitudes of 13.6 and 18.7, respectively. They are hard to photograph, especially as Triton stays relatively close to the planet. For success, one must use a relatively large telescope (300–500 mm) with a long focus ($F/D = 12$–20), and an exposure time of 6–15 min, on hypered TP 2415 film. The guiding, throughout the exposure, should be very exact.

One can also attempt to photograph the planet Pluto, against a background of stars, in order to record its displacement over a period of time. Its magnitude is never less than 14.9, and requires an exposure time of several minutes, at the prime focus of a 250–400 mm telescope. To be certain of

finding the planet against a starfield, it is necessary to take several photographs, separated by at least a week in time.

4.5 High resolution astrophotography in planetary, lunar and solar research

The contribution which amateurs can bring to various fields of astronomical research is a very controversial subject. For some, in certain fields, a real cooperation between amateur and professional is possible, and even desirable. The IAU is interested in this, and research programmes suggested by amateurs have been approved by the committee which coordinates the use of the Hubble Space Telescope. On the other hand, numerous professionals and non-professionals think that the amateurs can make but an insignificant contribution, since modern astronomy is becoming a more and more complex science, requiring extremely costly instrumentation and a more and more exhaustive background knowledge.

At the beginning of astronomical observation, there was no clear barrier between amateurs and professionals. The instruments had not yet assumed gigantic proportions and the techniques were still of a relatively rudimentary nature. The discovery of comets or asteroids, measurement of

Figure 4.46. Saturn viewed by the amateur. *1. By Günther Nemec, 1970 October 20 at 02.50 UT, with his 200 mm OG. 2. By G. Viscardy, with his 520 mm Cassegrain. 3. By G. Thérin, 1992 August 7 at 00.02 UT, with his 225 mm Takahashi Schmidt–Cassegrain; F/D = 95; W8 yellow filter; 5.5 s on hypersensitised TP 2415 film. 4. A rare photograph obtained by G. Nemec, 1966 September 12 at 00.47 UT: Saturn with its rings viewed edge-on, and the transit of Titan and its shadow (diameter 0".63), photographed with his 200 mm refractor; F/D = 120; 12 s on 17/10 DIN film. 5. 1975 December 24 at 01.31 UT, by the author, with the 106 cm Cassegrain of the Pic du Midi Observatory; F/D = 92; 4 s on Ilford Pan F film; developed in D19. 6. 1991 September 29 at 01.21 UT, by D. Parker, with his 410 mm Newtonian, at F/D = 27, and his CCD camera, 1.30 s followed by image processing. (On this superb image one can see various small white spots in the Equatorial Zone, remnants of the Great White Spot outburst of the previous year.) South is uppermost in Nos. 1, 2, 4 and 5; north is uppermost in Nos. 3 and 6.*

double stars and observations of planetary surfaces were often the fruitful outcome of the work of distinguished amateurs. It was above all in the exacting field of planetary observation that the amateurs seemed to enjoy great success. Atmospheric turbulence favours small instruments; all those who had keen sight and a talent for drawing could hope to make a contribution towards lunar or martian cartography, or to the study of the clouds of Jupiter by establishing the rotation periods of the various belts and zones. That is why amateurs such as Stanley Williams, Flammarion, Denning, Phillips, Comas-Solas, Dawes, Trouvelot, Antoniadi, Fournier, Peek, Saheki, etc., produced works of significant value. From 1880 to 1940, owing to the problems which faced professional astronomers in obtaining good lunar and planetary photographs, amateurs were able to play a considerable role in the evolution of our knowledge of the surfaces of the Moon, Mars and Jupiter.

After World War II, the quality of photographic images made appreciable progress, above all through the influence of B. Lyot and his team, and later through the work of A. Dollfus, Slipher, Smith and Reese, etc. At the same time, the number of amateur planetary observers notably increased but the quality was not in keeping with the quantity. In a general way, planetary drawing had received a mortal blow as a result of the excesses of Schiaparelli, Cerulli, Lowell and Douglass. In spite of the excellent results obtained by Barnard, Maggini and, above all, the great Antoniadi, A. Dollfus, G. de Vaucouleurs and Focas, the planetary drawing could no longer be considered as a valuable scientific document. Although the photographic resolution of the epoch 1936–40 was mediocre, the photographs obtained at the great observatories at least had an objectivity which few drawings could rival. As soon as high resolution photographs became more common, through the influence of the observatories of Pic du Midi, Flagstaff, New Mexico, etc., amateur planetary drawings lost all interest and were rapidly becoming unpublishable in the astronomical research journals.

Everywhere amateurs had worked with tenacity and courage: the Mars, Jupiter and Saturn Sections of the BAA (UK) have produced many interesting studies, followed by those of the ALPO (USA) and the Commission des Surfaces Planétaires (CSP) of the SAF (France), together with other groups of lesser importance. Although the founder member of the French CSP, then

Secretary and finally its President, I was never under any illusions about the real value of our publications, in amateur journals. Conversations with my professional friends A. Dollfus (who created and later presided over the CSP of the SAF) and G. de Vaucouleurs (who was the Secretary of the Mars Commission of the SAF, which preceded the CSP) convinced me that our *Travaux* were of a scientific nature and not devoid of a certain interest, but on the whole unacknowledged, being rather a source of discovery of young talented observers who might later aspire to becoming professional planetologists. The reality, however, is quite different: planetary research finally took a real upturn. After 1950 every means was employed to make rapid advances in this field of knowledge. High resolution photography was to compel recognition of the work of Lyot, Camichel, Gentily, Dollfus, Guérin, Smith, Reese, Slipher, Martin, Kuiper with his numerous co-workers, Kopal with his own team, and so many others. The famous 'International Planetary Patrol Program', subsidised by NASA, obtained tens of thousands of photographs, while eventually space probes photographing the planets at point-blank range have sent us an incredible number of images of Mercury, Venus, Mars, Jupiter, Saturn, Uranus, Neptune and their satellites.

In a relatively short time, planetary astronomy, a much neglected field in the 1940s, assumed great importance. What interest can there be, under these conditions, in the modest planetary observations of amateurs?

The discovery of the four-day retrograde rotation of the atmosphere of Venus is the wonderful modern discovery of the amateur C. Boyer, by means of high-resolution photography, but this case is unique in the history of planetology (see p. 139; Boyer, 1973 and Dragesco, 1979b). (Fig. 4.47)

In the meantime, however, amateurs, too, have made remarkable progress: excellent astrophotographers have started to provide high quality planetary documents in sufficient numbers to allow the study of atmospheric phenomena over very long periods, nearly on a day-to-day basis. The pioneers were, above all, G. Viscardy and D. Parker, followed initially by myself (though for a few years only), then by I. Miyazaki, with useful support from T. Platt. The observational reports by the coordinators of national organisations are becoming more and more complete, and the atmospheric phenomena of Mars and Jupiter can be studied quantitatively,

thanks to precise measurements from high resolution photographs. The sort of reports appearing in the *Journal of the British Astronomical Association*, the *Strolling Astronomer* or *l'Astronomie* (and later in *Pulsar*), being of a good scientific standard, bear a strange resemblance to the publications of the astronomers of New Mexico State University Observatory between 1960 and 1969. The only 'amateur' characteristic of such publications resides in their use of drawings in addition to more objective photographs. Visual observations should in my view be excluded without mercy, following the deplorable adventure of the Martian canals, and the uncertainty which affects all lunar drawings, compared with modern photographic documents, notably Orbiter imagery.* It is also regrettable that, quite often, the same photographs are chosen for illustration by the reporters of different organisations, so that their accounts of the same apparition are rather similar. I have always thought that the only way out of this rut, for amateur planetary astronomy, is to abandon the separate reports of national organisations and to publish a single combined quantitative report, in English, based only on the best photographs (whether amateur or professional), summarising the dynamical data pertinent to each apparition of Mars or Jupiter. Co-authored by a group of national recorders and well-known photographers, a single such report should be published in a research journal (*Icarus*, for example), and could therefore play a modest but significant role in the field of planetary astronomy.

For about a dozen years, amateur planetary observers were encouraged by some professionals, who maintained that planetary photography had been abandoned by NASA and by the specialist observatories. It is therefore now for the best amateurs to subscribe to the idea of providing, for the astronomical community, good-quality documents for following the yearly changes in the atmospheres of Mars and Jupiter, even Saturn (given that the Great White Spot of 1990 was

---

* The translator must differ from this view: the value of visual work is frequently to be found in the recognition of new phenomena, such as a new outbreak of the SEB of Jupiter, a Great White Spot on Saturn or a dust storm on Mars. If photographers are to compete in this area, they, too, must be as conversant with the visual appearance of the planet in the eyepiece as is the long-time visual observer at the moment of exposure. Drawing a planet is also an excellent way of training the eye, and the best visual observers can produce drawings considerably better than photographs obtained under average or poor conditions. The appearance of visual drawings in such reports by amateur organisations therefore need not, in my view, require any justification! (RJM)

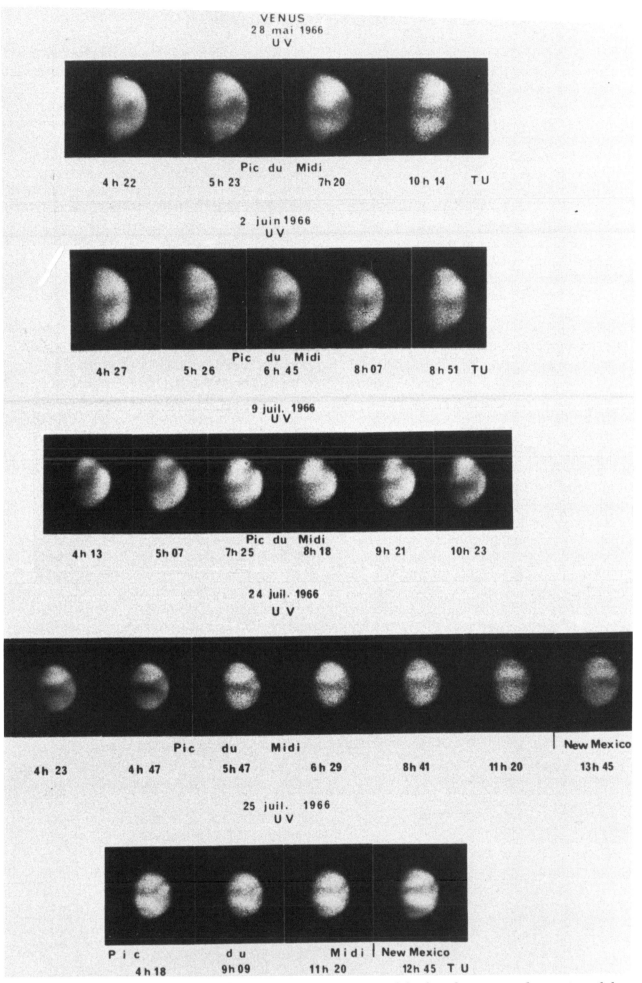

Figure 4.47. Some images of the planet Venus, in UV light, obtained at the observatories of Pic du Midi and New Mexico, following the discovery, by C. Boyer, of the four-day retrograde rotation of the atmosphere of that planet.

discovered and reported to the astronomical world by amateurs). In reality the professionals continue to take planetary photographs. With the assistance of NASA and the National Geographic Society, small programmes of planetary photography are always in operation, either at New Mexico, on Hawaii, at Catalina or at Pic du Midi. There is no longer the flood of documents of the years of 1969–70, but excellent photographs continue to be taken and studied. Even better, the 'International Jupiter Watch' was set up, which seems to be permanent and constitutes a very active research group (whose essential requirements completely surpass those of amateurs). Furthermore, an 'International Mars Watch' was created, with the help of the Planetary Society, whose objectives were more modest.*

In conclusion, I would say that there is still a small chance of doing useful work for the amateur planetary worker, in obtaining high quality documents (photographs, CCD images), with telescopes of 400 mm aperture or more, in favourable sites, and on condition that the quantity is on a par with the quality. Recent CCD images of Donald Parker show that he is on the right track, successfully photographing Saturn (1991) with very high resolution, and recording several small white equatorial clouds, which escaped the majority of telescopes. I. Miyazaki is going over to CCD work too, and the quality of his results will reach even greater heights. These two astrophotographers produce work of undisputed quality, and we can consider them to be valuable observational astronomers. But they are only two amateurs among more than 300 000 world-wide!

Can we speak, under these conditions, of 'an important cooperation between amateurs and professionals in the field of planetology'? In the

field of selenography practically everything has been accomplished by the spaceprobes (Apollo and Orbiter). Even if, in the case of the crater Fracastorius, G. Viscardy was able to discover some topographic details which escaped Orbiter 4 (see p. 101), the amateur, even the excellent astrophotographer, can scarcely hope to bring these new facts to light unless he is seriously involved in the 'Luna Incognita' programme of the ALPO (directed by Dr John Westfall).

As far as the Sun is concerned, there are throughout the world very many well-equipped solar observatories which enjoy excellent atmospheric conditions. The Sun is therefore photographed very often, both in white light and in Hα. The amateur, even one reaching the heights attained by W. Lille, cannot claim to be doing useful work. Some have proposed a low resolution solar photography programme. Thousands of documents have been obtained, over many years, but the results have been most deceptive. It is therefore very difficult to design truly useful solar observing programmes which are open to the amateur.

4.6 Conclusion

High resolution astrophotography is the most difficult branch of the photography of celestial objects and very few amateurs have succeeded in it. We hope that their number will increase in the years to come. Obtaining a high resolution photograph is very arduous but brings with it great satisfaction, as, in order to succeed, one must struggle continually, both with the hazards of atmospheric turbulence and with instrumental and photographic problems. To my knowledge, there are hardly more than about 20 real experts in the field of high resolution among the thousands of amateurs who photograph the sky. Video is replacing photography. With the popularisation of the CCD receptor, the recording of finer and finer lunar and planetary details will become much easier. This book is quite probably the last to be devoted to high resolution obtained by the methods of photography.

---

* It is regrettable that some earlier projects, such as the 'International Jupiter Voyager Telescope Observation Programme', or IJVTOP of 1978–9, though well supported by a great many amateur observers, did not lead to much in the way of published results, with little or no feedback to the contributors. Such projects, if they are to be run at all, must be properly organised and adequately funded. (RJM)

Part 3

5

Short biographies of the specialists in high resolution photography

Specialists in high resolution photography are extremely rare. For the preparation of this book I studied the literature over more than 16 months in order to try to contact the best Solar System photographers. I first consulted the most important Western astronomical magazines: *Sky and Telescope, Astronomy, Astronomy Now, Ciel et Espace, Sterne und Weltraum,* the *Journal of the British Astronomical Association, Strolling Astronomer, The Astrograph, Pulsar, Astro-Ciel, Orion,* etc. I next consulted all the recent works on astrophotography (Covington; Dobbins *et al.;* Martinez; Gordon, Wallis & Provin; Bourge *et al.;* etc.). Finally, I entered into correspondence with the major astronomical associations in the Western world. To conclude, after having written about 100 letters, I was able to contact some 20 amateurs with a close acquaintance with high resolution astrophotography. I trust that those I was not able to contact will forgive me for having thus neglected them.

The exchange of correspondence and the examination of the photographs that I received have convinced me that the good high resolution astrophotographer is a rare bird! Therefore, I thought it would be interesting for readers to become acquainted with our exceptional colleagues, from short biographies of those who seem to me to be the most interesting. I have first paid homage to the pioneers, who showed the way, as we owe them the most. I grouped the experts in high resolution astrophotography into three categories: the pioneers, today's experts and tomorrow's experts. These ten people, selected from the tens of thousands of active amateurs, can be regarded as being exceptional.

5.1 The pioneers

5.1.1 Charles Boyer: discoverer and pathfinder

C. Boyer (Fig. 5.1:1) was born in Toulouse on 1911 September 28. Passionately interested in wireless sets, he built his first radio at the age of 15 and became an experienced radio amateur. He was also fervently interested in motorcycling.

After his legal studies at university, he became an examining magistrate and, at the age of 28, entered into collaboration in radio work with Henri Camichel, an astronomer at the Pic du Midi Observatory. After the war Boyer was named examining magistrate at Auch. The influence of H. Camichel on the one hand, and reading Camille Flammarion's *l'Astronomie Populaire* (*Popular Astronomy*) on the other, led to the construction of his first small (45 mm) refractor. Deciding to pursue astronomy, Boyer made himself follow private courses in higher mathematics twice a week for 4 years. Having made a 150 mm altazimuth Newtonian, he took up a systematic study of atmospheric turbulence, following the scale of A. Danjon, for comparing his results with those obtained by his friend H. Camichel, who was making the same measurements at Pic du Midi. Boyer established beyond doubt that the quality of the images at sea level was not comparable with that which could be obtained at Pic du Midi. Having learnt to fly, he became President of the Aero-club of Auch. Before long, however, the magistrate, pilot and astronomer began an overseas career. Named as President of the Bench (or Law Court) at Cotonou (then Dahomey), he installed there a 200 mm Newtonian reflector and established that the 'seeing' was quite exceptional: he published several articles on this subject. After he became President of the Bench at Brazzaville in 1955, he constructed a new 250 mm Newtonian, provided with an optical window. Although his telescope was mounted as an altazimuth, he conceived an ingenious system for moving his camera equipment in the image plane and was thus able to take, automatically, a large number of Solar System images. Little by little, he mastered the techniques of high resolution photography of the Moon, Mars and Jupiter.

Then, on the advice of H. Camichel, he took up the systematic photography of Venus in blue-violet light. Indeed, since the work by the American Ross in 1927, it had been possible to record dark markings upon the planet by ultraviolet photography, but Ross had not been

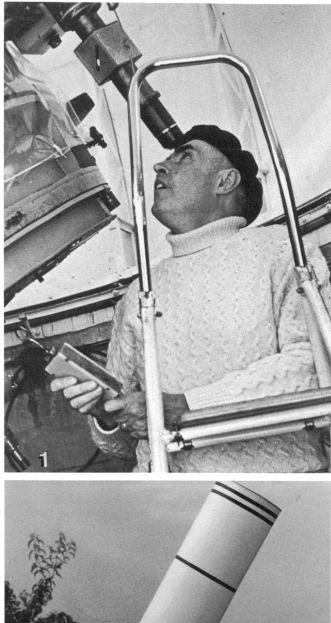

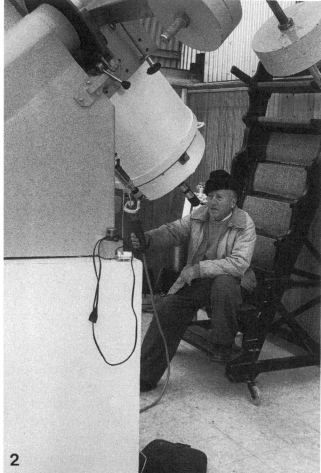

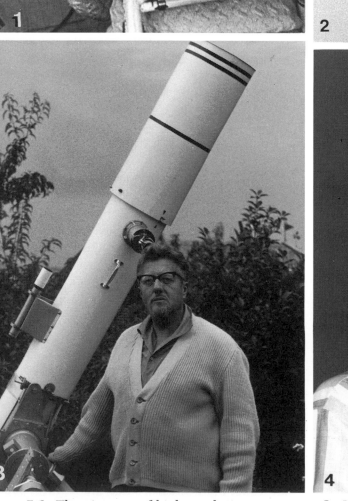

Figure 5.1. The pioneers of high-resolution astrophotography. 1. *Charles Boyer at the eyepiece of the 106 cm telescope at Pic du Midi.* 2. *Georges Viscardy beside his 520 mm Cassegrain telescope at* *St. Martin de Peille. 3. Günther Nemec and his famous 200 mm refractor in the Munich region.* 4. *Donald C. Parker at the eyepiece of his 310 mm Newtonian, at Coral Gables (Miami, Florida).*

able to determine the rotation period of the markings (at that time we were still assuming a synchronous rotation of 225 days). At Pic du Midi, Camichel had himself obtained UV photos of Venus which showed the dark bands, which seemingly indicated a rapid rotation of the markings, incompatible with the synchronous rotation. Boyer then undertook to photograph Venus every day in August and September 1957 (in the middle of the dry season of Brazzaville), with the aid of a violet filter, on high contrast film. Thanks to his keenness and to the excellent 'seeing' of the site, Boyer then discovered that the markings which he photographed were moving rapidly, indicating a retrograde rotation of about 4 days (and not 225!). Of course, this 'discovery' of a simple amateur was regarded with scepticism by many. Happily, Boyer was encouraged in his work by several professionals such as H. Camichel, J. Texereau, P. Guérin and J.-C. Pecker. Cautious, Boyer deposited, after 1957 September, a sealed envelope with the French Academy of Sciences. It was not until 1962 that A. Dollfus, convinced by J. Rösch and J.-C. Pecker, began a world-wide campaign to photograph Venus in the ultraviolet, with the cooperation of the observatories of Lick, New Mexico, Pic du Midi and several others. The photographs obtained confirmed the 4 day rotation in the retrograde sense (though some uncertainties existed, owing to the presence of similar dark formations, but situated at 90° from each other, the famous 'Y' and 'phi' (ϕ). From 1964, Guinot and Feissel were to confirm the rotation period by an altogether different technique: by measuring the radial velocities on Venus at different points of the limb.

In 1963, Boyer retired and was able to devote all his time to his Venus research. He became a professional astronomer, and the observatories of Meudon and Pic du Midi opened their doors to him. From July 1966 onwards, he obtained an excellent series of photos, which showed the two formations Y and phi. Reviewing with P. Guérin the existing collection of previous photographs he reconfirmed, once again, the retrograde 4 day rotation, which very obviously could refer only to the gaseous envelope of the planet (as Evans had proved, from radar echoes, that the surface of the planet rotated in 244 days). Although Boyer brought forward all the necessary explanations in a series of papers (some in English, with R. E. Newell or J. Nikander), the majority of American astronomers would not accept the conclusions of the French 'ex-amateur' until 1974, when the Mariner 10 probe provided irrefutable proof of his discovery. Boyer was to keep up his work, to make use of the best computers, and to show that the atmospheric movements of Venus speeded up as the clouds moved further onto the sunlit hemisphere, becoming most rapid in the evening. In 1986, he was to attempt to give an explanation of the variations in the speeds of the clouds throughout a Venusian diurnal cycle, but was not able to confirm his work, owing to serious illness. He died in Toulouse in 1989.

Boyer was the most illustrious of all the amateurs of our time and one of the few to become a member of the IAU (Commission 16). Between 1969 and 1974 he was also in charge of the Centre of Planetary Documentation at Meudon Observatory. I have the honour of having known him well and having been able to work at his side at Meudon and at Pic du Midi. He was an extraordinary man, very complex, fascinating, enthusiastic and lucid at the same time, a man the like of whom we meet only rarely. His discoveries being based on high resolution photography, Boyer will always remain the master of that art and the pathfinder for us all.

I thank J. Rösch (1990) and Mme J. Hecquet (1990), whose writings have given me the basis of these lines; also M. Futaully, who managed to provide me with photographs of Boyer and originals of some of his photos of the Venusian clouds.

5.1.2 Georges Viscardy: the relentless pioneer

Viscardy (Fig. 5.1:2) began to be interested in astronomy when he was less than 16 years old. As for many French astronomers of a certain age, *Popular Astronomy*, by C. Flammarion was the bedside book of his youth. Like the majority of us, Viscardy built, in 1934, a refracting telescope from spectacle lenses and cardboard tubes. His enthusiasm helped him, from the beginning of 1945, to build a dome and to try out a 110 mm refractor. From this time, he was involved in the intensive search for a favourable site and concluded it by discovering an exceptional microclimate with seeing sometimes better than 0".10. So, in 1953, Viscardy installed the first dome at St. Martin-de-Peille, at an altitude of 730 m, above Monte Carlo. At this time, all the material had to be carried up the mountain by mules. The dome tested out a Newtonian reflecting telescope, with mirror by A. Couder and mounting by Manent. A fine little darkroom adjoined the dome. From 1960, Viscardy enlarged the observatory, adding a sitting room, a conference room and a library, while a road paid for by himself connected the group of buildings. It was from that same year that Viscardy commenced his work in high resolution lunar and planetary photography. He created the Franco-Monegasque Astronomical Association and installed in his observatory many small instruments allowing for social evenings of stargazing.

From 1964, Viscardy took an average of 4000 photographs of each planet at each opposition with the 32 cm telescope. Little by little, true pioneer that he was, he refined his planetary photographic techniques and succeeded in filming the first complete rotation of the planet Jupiter, an exhausting task: 60 exposures per hour, for 10 h, on 1966 November 28, in the course of a single night! So, 600 photographs had to be rephotographed with the aid of a 16 mm ciné-camera, frame by frame, with precision alignment. The film was presented by Dr Focas and Viscardy during a seminar at Nice observatory. It was not until 1967 that Focas succeeded in refilming the rotation of Jupiter, at Flagstaff, but with the aid of photographs taken over two consecutive nights. Over more than 20 years, Viscardy would send his many planetary photographs to the Commission of Planetary Surfaces of the SAF or to other groups of planetary observers (notably the BAA). It was not until 1972 that Viscardy was able to install his 520 mm Newtonian–Cassegrain telescope. It was with the aid of this instrument, over the course of 3 years of strenuous work, that he obtained some 60 000 images which were to lead to the publication of his famous *Atlas-guide photographique de la Lune* (*Photographic Lunar Atlas and Guide*). This monumental work forms one of the fundamental documents we have for current selenographic information.

For more than half a century he has devoted himself to the pursuit of astronomy, in painful isolation, exposed to local incomprehension, without ever having received the least help or subsidy. However, in the words of Viscardy himself: 'We cannot be happier if, in the evening of our lives, we realise the dreams of our youth.' Sadly, for health reasons, Viscardy has now given up his astronomical activities.

I have the great privilege of knowing Georges Viscardy well. He is a very dear friend and all those who have approached him have been amazed at the enthusiasm he exudes and by his incredible tenacity. He is in other respects the simplest and most modest of men, and also the best friend in the world, whose generosity and patience have always been endless.

5.1.3 Günther Nemec: ace Solar System photographer

Günther Nemec (Fig. 5.1:3), born in 1921, of Czechoslovakian origin, lived in the town of Brunn. At the end of World War II he was obliged to settle in Munich, Bavaria. Becoming a professional photographer, Günther began to be interested in astronomy as a hobby about 1960, at which time he owned only a small refractor. In a few years the modest amateur became an expert in mechanical matters applied to astronomical instruments, and was quickly able to install an open-air observatory. Nemec did not want a dome, thinking with good reason that the images are more stable when a telescope is sited in the open. From 1965, he put into action his famous 200 mm refractor, with $F/D = 20$, giving a focal length of 4 m! The objective was a Fraunhöfer-type doublet, polished by Lichtenknecker. This exceptional objective was mounted as a refracto-reflector (following Schaer), a sort of folded instrument of reasonable length, permitting observation in a comfortable manner. Nemec was proud of the equatorial mounting, which he himself had designed and built, as he provided it with very precise movements allowing the Moon to be followed as accurately in RA as in declination. During the construction of his equatorial, Nemec was designing original mechanisms for which he obtained the patent rights, which he later gave up to the telescope makers.

Thanks to his exceptional refractor, to a rigorous photographic technique and to an incredible keenness, Nemec obtained, from 1969, results which were much superior to anything which had been published at that time in amateur astronomical magazines. Certain of his photographs were so detailed that many amateurs had difficulty in believing that they could have been obtained with an instrument of only 20 cm aperture. In 1970 Nemec was thus more than 10 years in advance of the majority of the other high resolution photographers of his time.

For a good decade, the lunar and planetary photographs of Nemec were considered unbeatable. He published his photos in numerous astronomical magazines, and they were accompanied by technical articles. It must not be forgotten that in his time the 'miracle' film TP 2415 did not exist, Günther being obliged to work with the less sensitive and more grainy emulsions of the day, Adox KB14 and KB17, notably, which he developed to perfection. Some present-day amateurs have been able to reach a resolution a little superior to that obtained by Nemec (about 0″.30 to 0″.40), but it is nearly a quarter of a century later! The results of Nemec remain, even today, among the best.

Before long, unfortunately, Nemec gave up practising astronomy at night. He was burgled on several occasions, as he lived in complete isolation on the outskirts of Munich (in order to enjoy a less polluted sky than in the centre). Being so very vulnerable, Nemec feared for his life (a well-known German amateur, a Jupiter specialist, had been murdered, at night, on the streets of Munich), and started to install more and more complex alarm systems. Eventually, Nemec, whose health had been weakened by many nights spent in photographing the sky, said that it was really too stupid having a hobby which made one live in fear of being killed, at any moment, by a criminal. Abandoning night-time work he continued for some time to take solar photographs in Hα light with a coronograph designed and built by himself, and in white light with the help of the 200 mm refractor. Between 1962 and 1972 he exposed some 60 000 negatives.

In 1972 Nemec decided to sell everything, and since that time astrophotography has ceased to exist for him. Over several years, he has studied microscopy and histological and bacteriological techniques. He developed a precision microtome, which enabled him to slice up thousands of microscopical preparations and to obtain superb photomicrographs. Then he became a semiprofessional film-maker, producing reports and documentaries for German television.

Günther Nemec is still extremely active. The great pioneer of high resolution astrophotography will always remain in our thoughts.

5.2 Today's experts

5.2.1 Donald C. Parker: planetologist

At the age of only 7, Donald Parker (Fig. 5.1:4) became interested in astronomy in 1946, through the influence of science fiction writings and children's books. Living near Chicago, he had access to the Adler Planetarium, where he was to meet other amateurs who made their own telescopes. Having constructed a refractor equipped with a simple lens, he was offered an achromatic telescope with a 35 mm objective by his father. Donald also made a small 75 mm Newtonian reflector, then attempted – from 1953 – to polish optical glass. He made a 150 mm

mirror ($F/D = 8$), but unfortunately, the 15-year-old was not able to obtain a successful parabolic surface. A year later, thanks to his grandmother, Donald was able to buy a good 200 mm mirror ($F/D = 8$) polished by Tom Cave. With his father's help he succeeded in putting together a good Newtonian reflector, with which he immediately began serious observations of Mars, Jupiter and Saturn. Spurred on by *Sky and Telescope* articles by Cave, Capen and Haas, Parker first undertook his systematic study of Mars during the famous opposition of 1956. Unfortunately, owing to the no less famous great dust storm which took place that same year, the young man was able to see but little of the Martian surface details.

From 1960 to 1970 Donald Parker neglected astronomy, as he was necessarily completely absorbed by his medical studies and then by his service in the US Navy. Between times he became a married man, with a family. It was not until 1972, now settled in Miami (Florida), that he was able to take up his astronomical work again, with his 200 mm telescope made 18 years earlier! So, he was able to study Mars in 1973 and to appreciate the fine images of the sky from Florida. Amazed at the quality of the images, Donald joined the ALPO at once, and put himself to work observing and drawing Jupiter and Saturn. During the 1975–6 opposition of Mars, Donald sent his drawings to the ALPO Mars Recorder, Charles ('Chick') Capen. Capen guided the young beginner and the two men became friends. Capen, a professional astronomer, had a great influence on Parker. It was he who was to encourage Donald to start out in astronomical photography and to guide him in this work (which he understood very well). At Capen's prompting, the ALPO created an 'International Planetary Patrol' which was very successful: in 1988 more than 300 observers from 22 different countries sent in some 7000 observations.

In 1978 Parker made his 300 mm telescope, equipped with an excellent $F/D = 6.5$ mirror by R. Fagin. It was with this instrument that Donald obtained numerous micrometrical measurements of the martian polar cap and so began his ongoing investigations into possible short-term variations in the martian polar climate.

Donald Parker has preferred to construct his own telescope mountings. The results have shown that he has succeeded better than most of the commercial suppliers. Donald also thinks that one should try to use a telescope with as large an aperture as possible for the quality of the site (at

Coral Gables, Miami, he has come to use the full aperture of his present 400 mm telescope). The large apertures give real benefit when it is a question of photography and colour filter work.

With his 300 mm Newtonian Donald Parker obtained excellent photographs of Mars and Jupiter, among the best in the world, on a par with those of G. Viscardy and G. Nemec. He rapidly became a renowned international astrophotographer, and a Mars Recorder for the ALPO. Since 1980, in collaboration with his friend and colleague Jeff Beish of Miami, Parker has published interesting contributions about the martian climate. The quality of his photographs of Mars and Jupiter improved continually, and some thousands of his images were to be published world-wide in astronomical magazines. With his 300 mm Newtonian, Parker was often obtaining images as good as those being produced by professional astronomers using powerful telescopes on mountain sites (*Sky and Telescope*, 1986). In 1986 he was sent to the Cerro Tololo Observatory to photograph Mars on behalf of the Lowell Observatory, University of Missouri, and the National Geographical Society.

In the course of his work, Parker had the opportunity to be co-discoverer of a Jovian SEB disturbance in 1979, and was the first to photograph a major new atmospheric eruption on that planet, in July 1985. At the end of 1988 Donald Parker was using a new 42 cm Newtonian telescope, which allowed him to obtain even more detailed planetary photographs. During the last 20 years, Don has produced more than 10 000 images of the Solar System. In 1991 he abandoned traditional photography in order to equip his telescope with a CCD and, with the collaboration of Richard Berry, he was quickly able to obtain images with a resolution scarcely less than that obtained traditionally at those observatories specialising in planetary observation. After having collaborated in the publication of a book, *Introduction to Observing and Photographing the Solar System*, he went on to work on a publication with R. Berry and J. Newton dealing with the acquisition and processing of CCD images. Over the past 15 years, Parker has authored or co-authored over 100 papers on the Solar System and on planetary photography. The majority of these articles were published in amateur magazines, but some appeared in professional journals such as *Science*, *Nature*, *Icarus*, *The Astronomical Journal* and the *Journal of Geophysical Research*.

The enthusiastic work of D. Parker, the excellent site in which he lives, and the quality of his instruments and techniques place him in the front rank of world specialists in planetary photography. For several years, his photographs have been used by the Planetary Research Centre, Flagstaff, or by the USGS for the production of high quality Mars maps. Parker was also one of the seven members of the imaging team of the Hubble Space Telescope, and also worked on the Dust Storm Project which was integrated in the research programme of the Mars Observer spacecraft, launched in 1992. Following the appearance of the Great White Spot of Saturn (September 1990), the Hubble Space Telescope imaging team asked Parker to photograph Saturn as often as possible with his CCD, to look for remnants of the Great White Spot. Parker's CCD images revealed residual equatorial white spots which permitted HST astronomers to time their observations so that these areas could be studied.

The indefatigable work of Donald C. Parker has brought him a number of distinctions: the Astronomical League award (1989); the ALPO Walter H. Haas award (1986); the Texas Star Party Lone Stargazer award; the WAA's Caroline Herschel Astronomy Project award; and the Astronomical League's Leslie C. Peltier award (1992), 'for his contributions to CCD imaging of the planets'. Dr Parker is a member of many astronomical associations and was a co-founder of the Institute for Planetary Research Observatories. Of course, he has given innumerable lectures across the United States.

Donald C. Parker is the best example of what can be achieved by a passionate and competent amateur who, having chosen a favourable observing site, is not only keen to collect useful data but also seeks to analyse his data and to draw from them interesting scientific conclusions. Also involved in planetological research, he is becoming by the force of circumstances, a free professional astronomer of world-wide reputation.

Donald is, in addition, a charming man and a true friend. His enthusiasm and his confidence are so evident that his example is salutary; planetary astronomy owes him much.

5.2.2 *Wolfgang Lille: The solar expert*

Wolfgang Lille (Fig. 5.2:1) was born in 1941 in the Hamburg region. As a Boy Scout, and travelling a lot, he loved to walk, with a compass and maps, in the forests and around the lakes of Sweden. At about 15, he began to be interested in

Figure 5.2. Some astrophotographers of great merit. *1. Wolfgang Lille, with his solar refractor at Stade (Germany). 2. Isao Miyazaki, ready to photograph Jupiter at Okinawa (Japan). 3. Terry* Platt *beside his quad-Schiefspiegler at Binfield (Great Britain). 4. Bernd Flach-Wilken (arrowed) with his friends at the observatory of Kichtrein (Thuringen, Germany).*

astronomy and, as a consequence of the deficiencies in the secondary education in his region, he visited scientific libraries, to study books on astronomy. In 1959 he constructed his first telescope: single lenses for the optics, mounted in a cardboard tube. The arrangement was quickly followed by a small 40 mm Kosmos refractor. At the same time Lille followed the courses in astronomy at the People's University and at the Hamburg Planetarium, where he was able – for the first time – to study the sky with a proper refractor: a 100 mm altazimuth by Carl Zeiss. Very quickly he was given the job of demonstrating the wonders of the night sky to the Planetarium's visitors. After his marriage (1962), he successfully continued in his profession as master goldsmith, as well as branching out into astronomy. 1964 was an important year for the 23-year-old. Lille became one of the founders of the 'Hamburger Sternfreunde' (today known as the 'Gesellschaft fur Volkstümliche Astronomie'), which Wolfgang has directed now for over 10 years. This association numbers 550 members and publishes a lively review *Sternkieker* (*Stargazer*). This same year, he made, successively, a 70 mm refractor, a 100 mm Schaer refractor ($F = 1500$ mm) and a 150 mm mirror, all polished by himself. In 1970 Lille also made a 150 mm Schiefspiegler ($F = 3000$ mm) provided with Lichtenknecker optics, and converted a 125 mm refractor into a coronograph.

Living in the little town of Stade on the Elbe (close to Hamburg), the talented goldsmith put all his skill and energy into the construction of more and more specialised astronomical instruments as, little by little, his principal interests turned towards solar work. Between 1980 and 1985 he was trying out various optical systems and taking up astrophotography in various fields: teleobjectives with Zeiss mirrors of 500–1000 mm focal lengths; a 200 mm Schmidt–Cassegrain; a Schmidt–Cassegrain camera; a 100 mm Vixen refractor ($F = 1500$ mm); and fluorite refractors (80/640 and 102/900). He even owns one of the first Starfire 200 mm refractors ($F = 3000$ mm).

The interest of Lille in solar studies involved the installation, in 1985 of a highly specialised instrument, provided with a simple ('chromat'!) 175 mm objective with a focal length of 2118 mm. To eliminate chromatic aberration of the simple objective, Lille uses interference filters; one of them, which has a 9 Å bandwidth, is centred in the green (5150 Å; 515 nm) and serves for the photography of sunspots and solar granulation. The excess light is eliminated by a Herschel wedge. Lille obtains very long focal lengths, 10–30 m ($F/D = 150$) by two methods: either with strong negative Barlow lenses, or with short-focus eyepieces. In spite of the long exposure times (almost 1/60 s, on TP 2415 film), he succeeds in obtaining incredibly fine solar photos, and regularly achieves a resolution of 0″.5 or even less (details of sunspot penumbral filaments, rice grains and their complex form and substructure!). The same objective allows him to photograph prominences, with the aid of a coronograph of his own making, provided with a 10 Å bandpass filter centred on the hydrogen alpha line at 6563 Å (656.3 nm). The focal lengths reach almost 11 m (exposure times between 1/25 s and 1/30 s, Kodak TP 2415 film). Lille also employs a hydrogen alpha Day Star 0.5 Å filter, which he uses with a focal length of 15 m, to record chromospheric structures (exposure time 1/125 s). The results are surprising: with an average resolution of 0″.9, some photos seem as if they were taken directly at the world's largest solar observatories. But Lille has not said the last word. Since 1990 he has installed, in place of the 175 mm, a simple 300 mm ('chromat') objective of 3700 mm focus made by Dany Cardoen. With this enormous refractor (mounted, in part, in the open air; see Fig. 4.2), used with his Herschel wedges, monochromatic filters and focal lengths up to 125 m, the resolution on the solar granulation seems to me to reach 0″.25! The German amateur has arrived at a standard hard to beat: his best images with the 'chromat' refractor are certainly worthy of comparison with those obtained at the world's best solar observatories. However, these results were obtained in an ordinary small town at sea level. How is such success explained? (I myself photographed the Sun for more than 30 years without ever having obtained a resolution better than about 0″.80, even from sites with low turbulence such as Cotonou.) The German Cord Hinrich-Jahn has already obtained, with the 200 mm refractor of the University of Hannover Observatory, sensational photographs of the solar granulation (reaching the theoretical resolution of the instrument), but Lille has gone further than any other amateur (and as far as many professionals). His success is partially explained by his 'non-conformist' technique: simple objective of large diameter, no frontal rejection filter, use of a large Herschel wedge coupled with monochromatic interference filters, and use of very long focal lengths. But the essential thing, to my mind, is the fact that Lille spends hundreds of hours at the binocular eyepiece of his telescope, searching without rest for the moments of atmospheric calm. He himself reckons (Lille, 1992) that the use of very long focal lengths (up to 20 m) is of no importance for 99.9% of the time but, during the rare favourable moments, he obtains an important increase in resolution (for the hundreds of hours spent waiting for the right instant). Lille's techniques are complex and original: they are not open to the beginner.

But Wolfgang is still quite young, and he will be surprising us again. He plans the construction of a new Schiefspiegler telescope of the Yolo-System type of 350 mm aperture (and perhaps even a 600 mm!). He is also a great traveller: solar eclipses (1961, in Italy, 1973, 1976 in Kenya, 1983 in Java and 1991 in Mexico); visits to observatories – Puimichel, Nice, St. Michel, Tenerife, La Palma, the great American institutions (notably solar ones); etc.

So, although lacking a scientific training, thanks to his enthusiasm, tenacity, originality and skilful technique, Wolfgang Lille has raised himself to the front rank in the world, in the most difficult field that exists for the amateur astrophotographer: high resolution solar work, in white light and in hydrogen alpha.

5.2.3 Isao Miyazaki: the unbeatable young planetary photographer

Born in Tokyo on 1961 January 20, Isao Miyazaki (Fig. 5.2:2) began to observe the heavens in 1973, i.e. at the age of only 12, with the aid of a small telescope: a 100 mm Newtonian. When he moved to the island of Okinawa, the young Isao specialised, from 1976 onwards, in observing and drawing the planet Jupiter, with the aid of an excellent 200 mm Newtonian, on an altazimuth mount. Blessed with an eagle eye and being an excellent artist, Miyazaki rapidly gained an international reputation. About 1982–3, though aged only 21, this skilful Japanese observer was already sending his drawings of Jupiter to various planetary observing groups. In the Commission des Surfaces Planétaires of the SAF we were able to compare Miyazaki's drawings with the photographs of our colleagues D. Parker and G. Viscardy, with those taken at Pic du Midi and with my own images obtained at Cotonou. The comparison showed that the drawings of Miyazaki were more detailed than the best photographs and of a very high accuracy.

The island of Okinawa is surrounded by

magnificent coral reefs which Miyazaki has been able to study since his years spent at university, and which were to lead to his obtaining a Diploma in Biology. Miyazaki is a quarantine inspector at the Naha Protection Station of the Ministry of Agriculture and Forestry. He married in 1987; his wife, like him, is passionately interested in astronomy.

In 1988 Miyazaki was finally able to acquire an exceptional telescope. The mirror, figured by I. Tasaka, measures 404 mm in diameter and the focal length is 240 cm. The mounting is on a very heavy German equatorial, which was conceived and built by one of Miyazaki's friends, the founder and director of the Nagato Optical Company. The telescope is of professional quality (Fig. 2.2:4); few amateurs aged 27 could dare to hope to own such an instrument. It is installed in the open air, on a large terrace, and enjoys good observing conditions.

Since the end of 1988 Isao Miyazaki has devoted himself exclusively to the photography of the planets Jupiter, Mars and Saturn. He obtains 200–400 high resolution photographs at each opposition. The majority of his photos are admirable, showing details as small as 0″.5 or even 0″.4; I measured, on one of the best images of Jupiter, a thin belt only 0″.3 wide. This resolution compares with that obtainable in the best images of the observatories of Catalina or Pic du Midi (by traditional photography).

The technique used by Miyazaki is very simple, and largely inspired by that of D. Parker. He employs long focal lengths ($F/D = 100–187$) by making use of short-focus eyepieces (5 mm or even 3.8 mm focal length). Under these conditions the optical paths are short and the assembly remains compact, despite the body of the Nikon 501 reflex camera (equipped with a film dater). The film used is Kodak TP 2415. The impressive weight of his telescope allows Miyazaki to make use of the shutter of the reflex camera without fear of troublesome vibrations. The exposed TP 2415 film is developed in Rodinal developer, dilution 1 : 100 for 12 min at 20 °C, or at dilution 1 : 50 for 14 min at 20 °C. In these conditions (large focal ratios and very moderate development times) the exposure times are necessarily long: 2–3 s for Mars without filter, or 15 s with violet filter; and 2.5–5 s for Jupiter without filter, or 5–7 s (with red filter). One has to give almost 15–20 s for Saturn. These long exposure times demonstrate, on the one hand, the excellence of the seeing of Okinawa (during the months of June to September) and, on the other hand, the quality of the drive of the telescope used. Unfortunately, on Okinawa island the storms of the monsoon season as well as cyclones quickly tear off the tarpaulins which cover the 404 mm telescope. Miyazaki eventually hopes to be able to house his reflector under a dome.

Isao Miyazaki is, in my view, the best amateur planetary photographer in the world: his finest photographs have the clarity of those obtained in the best observatories which still use photographic film. However, Isao has now turned to CCD work and the resolution of his images has reached new heights. The astronomical 'career' of Isao Miyazaki is appropriately dizzying. The young amateur who began his career in 1980 is becoming a planetologist, having co-authored research publications. His images are known throughout the world, and his recent photos of the Great White Spot of Saturn (1990) have been very useful. A wonderful future opens before Isao Miyazaki, a man with a passion for the planets.

5.2.4 Bernd Flach-Wilken: all-round astrophotographer

Born on 1952 July 31, Bernd Flach-Wilken (Fig. 5.2:3) became interested in astronomy at the age of 17. His first instrument was a 70 mm refractor. In 1971, at only 19, Flach-Wilken constructed his first reflecting telescope, a 150 mm Schiefspiegler, equipped with mirrors polished by Lichtenknecker. Furthermore, he became interested in lunar and planetary photography and obtained encouraging results. Unfortunately, between 1975 and 1983 the young Bernd had to abandon astronomy to study pharmacy at Bonn and Dusseldorf, eventually working for the Bundeswehr, in the field of analytical chemistry. Finally, having been able to buy a pharmacy in Westerwald, he had the chance, at the age of 31, of taking up astronomy again, after a long break. As mentioned already, Flach-Wilken began by building an excellent 300 mm Schiefspiegler telescope with the aid of Lichtenknecker mirrors. Admirably mounted as an equatorial, the instrument proved to be excellent and allowed Bernd to start taking very good lunar and planetary photographs again, attaining a resolution of $0''.30$.

Flach-Wilken has since become perhaps the most all-round German astrophotographer, being as successful in solar photography as he is in deep-sky work. He owns a Starfire 178 mm refractor, as well as a flat-field camera by Lichtenknecker (300 mm diameter, $F/D = 32$),

and has also been able to acquire an excellent Schmidt telescope (300 mm, $F/D = 1.7!$).

With the aid of his 178 mm refractor, the German photographer obtains superb photographs of solar prominences, thanks to a complementary coronograph specially produced by Baader for the Starfire. He is one of those rare amateurs who can obtain a resolution better than $1''$ in solar prominence photography. He also excels in deep-sky photography. Bernd Flach-Wilken is still young; a brilliant future is before him. He envisages mounting a high quality CCD detector on his 300 mm telescope and so attaining increased resolution. He also hopes to obtain wide-field photos, in the domain of the long wavelength of hydrogen alpha, with the aid of narrow-band filters, during his journeys through the Alps with his new Schmidt.

To conclude, I consider that Bernd Flach-Wilken is one of those rare all-round photographers who is brilliantly successful in nearly every field of astrophotography.

5.2.5 Terry Platt: a modern amateur who looks to the future

Terry Platt (Fig. 5.2:4) is employed as technical director of an electronics company in Maidenhead, England. He was born in 1946 March 30, and can trace the origins of his interest in astronomy back to 1955, when he received an encyclopaedia as a Christmas present. The book contained pictures of the Moon and planets and a diagram of the 200 in Hale reflector. Various abortive attempts to make a reflecting telescope based on a concave shaving mirror were rewarded when his parents gave him a good-quality 32 mm refractor. This was used on many nights to seek out objects listed in various library books, and Terry attempted to photograph the Moon by placing a cheap 120 roll film camera behind the eyepiece. The sky at his home in the small village of Gargrave in the West Riding of Yorkshire was exceptionally dark. A small telescope could give remarkably good results under such conditions, and this served to maintain Terry's enthusiasm.

The appearance of Comet Arend-Roland in 1957, combined with the start of the Space Age, confirmed his interest in the night sky, but he was also very enthusiastic about electronics, and made many radio receivers and transmitters from parts scavenged from old TV sets. The idea of using a TV camera to display the view through a telescope dawned upon Terry about this time, but he had to wait many years before the components were available and he had the necessary ability!

In 1961 he decided to try to build a reflecting telescope, and saved up enough money (£9 10s 6d or £9.55) to buy a 6 in (152 mm) $F/D = 8$ mirror from Brunnings in London. A lot of 'tin-bashing' and searching through scrap metal resulted in a fairly good telescope on a fork mount in his parent's garden in Yorkshire. He experimented with various films, made several telescope drives, tried his hand at mirror grinding and built radio telescope receivers during the next few years, during which time he attended grammar school in Skipton. The school acquired a 4 in (102 mm) refractor, and he became involved with the school astronomical society and joined the BAA.

He left school with A levels in Maths, Physics and Chemistry, and nearly started on a degree course at university. However, he decided to earn a living instead and joined Racal Electronics, in Bracknell, Berkshire. One of the firm's engineers had an old security TV camera, and Terry soon attached this to a 10 in (254 mm) Newtonian, with encouraging results. As Terry recalls, unknown to him at this time (1970), the first CCD prototypes were about to be made at Fairchild Electronics in the USA.

Terry married in 1969, and today has two children. His TV experiments continued until about 1980, when he obtained several silicon target vidicon tubes from a local surplus shop. They proved to be very suitable for astronomy, and Terry's first planetary images were secured with a 200 mm Newtonian and published in *The Astronomer*. To allow for long-exposure work, Terry developed a solid state 'frame grabber', which was coupled to the only computer that he could afford at the time: a Sinclair Spectrum! This gave good results on the Moon, Jupiter and Saturn, but longer exposures were still not very practical, owing to the lack of cooling of the vidicon tube. However, in 1986 Sony produced a relatively low-cost CCD imager for their portable camcorders (the ICX021) and one of these was purchased to replace the vidicon camera. The results were dramatic, and Terry has since developed a universal CCD system for both planets and deep-sky objects, allowing for enhancement of the images.

The telescope which Terry uses for planetary imaging is an off-axis 318 mm quadri-Schiefspiegler. This is effectively a tri-Schiefspiegler equipped with a supplementary plane mirror: see Fig. 2.4. The original Schiefspiegler of Anton Kutter offered the advantage of being free from a central obstruction, but his basic model, with two mirrors, is not easily adapted for making large telescopes. We must thank Richard Buchroeder of California for the idea of adopting a third concave mirror, allowing for good coma correction, even for large apertures: this is the tri-Schiefspiegler. The quadri-Schiefspiegler of Terry Platt consists of a 318 mm $F/D = 12$ primary mirror, of ellipsoidal shape, tilted at 3°, and a 150 mm secondary, convex and spherical, having the same radius of curvature as the primary and tilted at 6°. The third mirror, 150 mm in diameter, is concave and spherical, its radius of curvature being 45 m; it is tilted at 45°. A 100 mm plane mirror projects the final image in such a way as to make the final image comfortable to view (Fig. 2.4). This system works at $F/D = 20$ overall, but is normally used with a mobile Barlow lens at about $F/D = 40$ to adjust the image scale for planetary CCD work. The lack of a central obstruction gives a performance close to the theoretical limit, but experience has demonstrated that the seeing conditions in southern England are by far the most important factor!

The CCD which is in use at present is the Sony ICX027BL-6. This is an 'Interline Transfer' device with 500 pixels per line and 256 useful lines, the physical size being about 6.4×4.5 mm. Terry explains that Interline CCDs are not popular for astronomy, owing to the wasted area of the sensor occupied by the vertical transfer registers. However, the technology of the modern devices has advanced to the point where microscopic lenses on the CCD surface are used to concentrate the incoming light into the sensitive zones, and ultra-low dark-current pixels allow long exposures without cooling. The spectral response is very much better in the blue than that of most other CCDs, and a built-in electronic shutter allows exposures to be started and terminated at precise moments. When all the features are taken into account, the ICX027 is an excellent choice for amateur CCD work.

The overall system has evolved into a modular arrangement of a 'frame store' (to hold the TV image), camera driver box, and CCD camera head. A 5 m cable links the camera to the frame store and this is, in turn, linked to an IBM PC computer to allow disc storage and image processing. Planetary images are generally recorded with exposure times of about 0.5 s and 'downloaded' to the TV monitor in about 1.5 s. This rapid cycle time allows many images to be seen in almost real time during an observing session, and the best of these can be recorded on disk for processing the following day. The enhanced CCD images are then photographed from the monitor screen. Terry emphasises that the rapid assessment of the results is an enormous advantage over conventional photography, added to which one has the ability to enhance contrast and definition of the images.

Terry Platt is now producing CCD systems using the ICX027 for amateur use on a commercial basis, and hopes that this will promote the spread of high resolution CCD imaging in the amateur community. His CCD images are very good, the best we have seen from Great Britain. They are important scientific documents, which Terry has submitted to the BAA and similar organisations.

5.3 Tomorrow's experts

5.3.1 Christian Arsidi: the Moon lover

Christian Arsidi (Fig. 5.3:1) was born on 1951 December 13 in Paris. He became interested in

astronomy when 17 years old, and in the beginning used a telescope made from spectacle lenses, followed by a small 60 mm refractor. His passion for the sky increased further in 1969 at the time of Apollo 8. He acquired a 115 Newtonian (900 mm focus), which he used until 1978. Photographer by training, Arsidi first felt attracted to the study of astrophysics and cosmology. It was not until the age of 27 that he was to buy a 203 mm Schmidt–Cassegrain, with which he began astrophotography from the suburbs of Paris (La Celle – St. Cloud).

He was especially interested in the Moon and planets, and very soon was obtaining excellent lunar photographs with the C8, which were really skilful for the time, allowing for the imperfect drive of his telescope and the poor resolving power of contemporary films (Ilford FP4). A number of Arsidi's lunar photographs were published widely, as well as some of his excellent planetary images. Christian became a prizewinner of the Association Française d'Astronomie in 1979, receiving the

Figure 5.3. Two young astrophotographers. *1. Christian Arsidi with his portable Takahashi 250 mm Mewlon Cassegrain (Dall–Kirkham) telescope, on an EM 200 Takahashi mount and Meade tripod. Arsidi generally uses his 310 mm Cassegrain for very high resolution lunar photography. 2. Gérard Thérin with his Takahashi 225 mm Schmidt–Cassegrain on Vixen 106 mount and Meade tripod. This instrument is capable of photographing lunar details smaller than 0".30.*

Julien Saget award of the Société Astronomique de France in 1982. After brief experiments with a C11, Arsidi decided to invest in a fine 305 mm Cassegrain, of which the mirror – polished by R. Moser – had a focal ratio of $F/D = 4$ and a resultant focal length of 7 m.5. Christian designed the tube and left the construction to excellent craftsmen. The optics tube, which weighs no more than 25 kg, was installed on a heavy AE equatorial mount the drive of which was improved (with a precision of 5 μm on the worm-wheel). The combination is of an exceptionally high quality, since Arsidi has been able to record lunar details smaller than 0″.25.

Devoted to both optics and photography, Arsidi also discovered the pleasure of visual observation, in the course of practising high resolution photography with the aim of surpassing his own standards. Very meticulous, he always works with the telescope perfectly collimated and uses a motor for setting the camera shutter (in order to avoid touching the telescope). Arsidi makes use of the manual shutter method, wearing gloves to move the occulting disk in front of the telescope tube (in order that the heat given out by his hands should not degrade the seeing). He is located in a favourable spot as far as wind is concerned: because he has no observatory, his telescope is mounted in a courtyard and covered with a tarpaulin. He has always used the telescope in the Paris region.

Arsidi met G. Thérin in 1984, and they have often worked side by side. I think that Arsidi had a very good influence on his young friend.

Married with two children, Arsidi has become the French representative of the American firm Meade, and manages a superb shop for astronomical instruments ('Le chasseur d'étoiles'). He now has much less time for astrophotography. The Moon remains, as always, his preference. He finds that the variety of lunar landscapes is extraordinary, the angle of illumination, and therefore the appearance, changing constantly. Thanks to the miracle film TP 2415, new horizons open up before the high resolution amateur photographer. Christian Arsidi leaves nothing to chance, and pushes to the limit the logic of perfection. His lunar photographs are known to the whole amateur community (and have even been published in the USA!). By his advice and his example, Arsidi has had a considerable impact on all the young astrophotographers who aspire to his ideal: to obtain the perfect image, the perfect astrophoto at all costs.

5.3.2 Gérard Thérin: a talented youngster

Gérard Thérin (Fig. 5.3:2) was born on 1962 March 22 in Paris. After his secondary education and baccalaureat, he entered the Police Force. He is married, with a small son born in 1991.

Thérin's interest in astronomy began at the age of 12, when he observed the sky with a 115 mm Newtonian ($F = 900$ mm), a present from his mother. It was with this modest instrument that he became familiar with the Moon and planets. However, it was not until the end of 1981–2, i.e. when about 20 years old, that he began seriously to undertake astrophotography, having acquired several successive telescopes: a 150 mm Newtonian, then another of 200 mm, and finally a 203 mm Schmidt–Cassegrain (Celestron 8) of excellent quality, which enabled him to take successful photographs of quite exceptional value.

Although lacking from the beginning a scientific upbringing, Thérin was to benefit from two opportunities: first he was able to get an introduction to photography in the club at his school and rapidly became an experienced worker; second, he made the acquaintance (in about 1984) of C. Arsidi, who was already a well-known astrophotographer, and from him Thérin received much advice and practical help.

Living in a small apartment on the third floor and without an escalator, Thérin has, by necessity, become the owner of portable instruments. He has opted for a C8 tube, which he has carefully installed on a solid German Vixen 106 mounting (with 40 mm diameter axes, built-in polar alignment and a very precise stepping motor drive). In 1992 the C8 tube was replaced by a Takahashi TSC 225 Schmidt–Cassegrain tube of very high quality. With the two instruments Thérin has been able to obtain high resolution photographs of the Moon which are among the best in the world.

His approach in the field of astrophotography is very simple: he tries to get the best from his telescope by waiting for the best moments, on the focusing screen, when the turbulence becomes low and allows him to record details of incredible delicacy. His ultimate aim is to photograph all that is visible through the eyepiece. In order to improve the chances of getting good images, Thérin does not hesitate, during his holidays, to take his telescope to more favourable places, such as, for example, the Calern plateau (Côte d'Azur Observatory).

I consider that Gérard is one of those rare astrophotographers who could be considered to be all-rounders, and in spite of his youth he has already succeeded magnificently in solar, lunar, planetary and deep-sky photography. His achievement is all the greater, as he does not possess an observatory, not even a terrace, garden or balcony! Each time, he must carry, mount and align his telescope (for most of the time largely useless, as nights with very low turbulence are rare).

The lunar photographs of G. Thérin with his Celestron 8 have created a great stir since they became known. Many amateurs refused to believe in their authenticity (as at other times with G. Nemec's lunar photos). They would not admit that those images could be obtained with a C8 with a large central obstruction! In reality, the results which Gérard obtains are theoretically possible but difficult to achieve. Thérin has no special secret. He is an extremely meticulous operator, who neglects no detail (collimation,

temperature control, perfect polar alignment, etc.). His technique is very simple: thanks to two Barlow lenses ($\times 2$) and an adequate projection distance, he has chosen a focal ratio between $F/D = 60$ and $F/D = 80$. He uses a motorised Olympus OM1 camera body and his exposure times (on TP 2415 film) are between 0.8 s and 1.5 s. The exposure is given by a vibration-free Compur shutter, or by using the 'hat-trick' method. For good images, Thérin takes 150 exposures during just one session. The films are developed in HC-110 (dilution B) for 9–14 min at 20 °C. Enlarging onto paper is difficult (as the negatives are very contrasty), but Gérard has a great deal of darkroom experience. Just as successful in deep-sky photography with a small 100 mm Genesis refractor, as in the field of high resolution, Gérard Thérin represents one of the hopes of the astrophotography of tomorrow.

References and Bibliography

Acker, A. (1987). *Formes et couleurs dans l'Univers.* Paris: Masson.

Adams, A. (1948). *The Negative.* New York: New York Graphic Society.

Alter, D. (1968). *Lunar Atlas.* New York: Dover.

Baum, W. A. (1973). The International planetary patrol program: an assessment of the first three years. *Planet Space. Sci.,* **21**, 1511–19.

Baxter, W. M. (1963). *The Sun and the Amateur Astronomer.* London: Lutterworth Press.

Beish, J. (1992). IR-blocking paint for telescope and observatory. *Strolling Astr.,* **36**, 15.

Bell, L. (1922). *The Telescope.* New York: McGraw-Hill.

Berry, R. (1991). *Introduction to Astronomical Image Processing.* Richmond, VA: Willmann Bell.

Berry, R. (1992). *Choosing and Using a CCD Camera.* Richmond, VA: Willmann Bell.

Béruex, M. (1958). Contribution à la connaissance de l'atmosphère équatoriale, une année de radio-sondages à Léopoldville, Congo Belge. *Mém. Ac. Roy. Sciences Col.,* V (**5**).

Bourge, P., Dragesco, J. & Dargery, Y. (1979). *La photographie astronomique d'amateur.* Paris: P. Montel.

Boyer, C. (1954). La turbulence au Dahomey pendant la saison sèche 1953–1954. *L'Astronomie,* **68**, 396–7.

Boyer, C. (1958). Etude comparative de la turbulence au Pic-du-Midi, Cotonou, Brazzaville, Toulouse, Mont-Louis et Auch. *L'Astronomie,* **72**, 379.

Boyer, C. (1973). The 4 day rotation of the upper atmosphere of Venus. *Planet. Space Sci.,* **21**, 1559–61.

Boyer, C. & Camichel, H. (1961). Observations photographiques de la planète Vénus. *Ann. Astrophys.,* **24**, 531–5.

Boyer, C. & Guérin, P. (1966). Mise en évidence, par la photographie, d'une rotation rétrograde de Vénus en 4 jours. *C.R. Acad. Sci. Paris,* **263**, 258.

Boyer, C. & Newell, R. E. (1967). Ultraviolet photography and the radar cross-section of Venus in 1966. *Astr. J.,* **72**, 679–81.

Bray, R. J. & Loughead, R. E. (1964). *Sunspots.* London: Chapman & Hall.

Buil, C. (1989). *Astronomie CCD.* Toulouse: Soc. Astr. Pop.

Callatay, V. de (1962). *Atlas de la Lune.* Paris: A. de Visscher–Gauthier-Villars.

Capen, C. F. (1967). The amateur and solar system photography. *Popular Astronomy,* **61**, 6–12.

Capen, C. F. (1970). *Observational Patrol of Mars in Support of Mariner IV and VII.* Jet Propuls. Lab. Calif. Inst. Technol.

Clerc, L. P. (1939). *Structure et propriétés des couches photographiques.* Paris: Editions Revue d'Optique.

Cortesi, S. (1974). La qualité des images télescopiques à Locarno Monti. *Astronomische Mitteil.,* 334.

Coulman, C. E. (1969). A quantitative treatment of Solar seeing. *Solar Phys.,* **7**, 122–43.

Covington, M. A. (1985). *Astrophotography for the Amateur.* Cambridge: Cambridge University Press.

Curtin, D. & de Mayo, J. (1982). *The Darkroom Handbook.* London, New York: Van Nostrand.

Dall, H. E. (1938). Diffraction effects due to axial obstruction in telescopes. *J.B.A.A.,* **48**, 163.

Danjon, A. & Couder, A. (1935). *Lunettes et Telescopes.* Paris: Editions Revue d'Optique.

di Cicco, D. (1989). Celestron vs. Meade: An 8-inch showdown. Part 1: Optical tests and performance. *Sky and Telescope,* **78** (6), (Dec.) 576–82. [See also Part 2: Mechanical tests and performance, *ibid.,* **79** (1) (Jan.), 33–9.]

Difley, J. A. (1968). Two photographic developers for astronomical use. *Astr. J.,* **73**, 762–8.

Dijon, J., Dragesco, J. & Néel, R. (1987). Les surfaces planétaires. In Martinez, P., *Astronomie (Le guide de l'Observateur),* pp. 195–259. Toulouse: Soc. Astr. Pop.

Dobbins, T. A., Parker, D. C. & Capen, C. F. (1988). *Introduction to Observing and Photographing the Solar System.* Richmond, VA: Willmann Bell.

Dollfus, A. (1961). Visual and photographic studies of planets and satellites at the Pic-du-Midi. In Kuiper, G. and Middlehurst, B. A. (eds.) *Planets and Satellites,* pp. 534–71. Chicago: Chicago University Press.

Dragesco, J. (1969). La vision dans les instruments astronomiques et l'observation physique des surfaces planétaires. *L'Astronomie,* **83**, 355, 399, 439.

Dragesco, J. (1976). Essai de télescopes Schmidt–Cassegrain Celestron. *L'Astronomie,* **90**, 238–45.

Dragesco, J. (1978). Réflexions sur la photographie astronomique à haute résolution. *L'Astronomie,* **92**, 31–9.

Dragesco, J. (1979a). Osservazione visuale e fotografice delle superficie planetarie. *Giornale d'Astronomia,* **5**, 283–313.

Dragesco, J. (1979b). The real story of the discovery of the rotation in four days of Venus' atmosphere. *Strolling Astronomer,* **27**, 173–4.

Dragesco, J. (1984). High resolution lunar photography. *Astronomy*, Oct., 35–8.

Dragesco, J. (1986). High resolution views of Mars and Jupiter. *Sky and Telescope*, **77** (June), 680–1.

Dragesco, J. (1987a). Introduction à l'étude expérimentale du défaut de réciprocité des émulsions sensibles avant et après hypersensibilisation par l'hydrazote. *Pulsar*, No. 662, 193–5.

Dragesco, J. (1987b). Défaut de réciprocité des émulsions sensibles avant et après hypersensibilisation par l'hydrazote 1°. *Pulsar*, No. 663, 95–8.

Dragesco, J. (1987c). Reciprocity failure of photographic emulsions before and after hypersensibilisation by forming gas. *Astrophoto VII.* Orange County Astronomers. pp. 42–8.

Dragesco, J. (1988a). Défaut de réciprocité des émulsions sensibles avant et après hypersensitisation par l'hydrazote 3°. *Pulsar*, No. 666, 174–9.

Dragesco, J. (1988b). Reciprocity failure of photographic emulsions before and after hypersensitisation by forming gas: a densitometric study. *J.B.A.A.*, **98**, 249–54.

Dragesco, J. (1992). Pouvoir séparateur, résolution, définition, notions importantes, souvent mal comprises. *Astro-Ciel*, **42**, 27–8.

Dragesco, J. (1993). Deux dispositifs nouveaux pour la photographie à haute résolution. *Pulsar*, No. 697, 106–10.

Dragesco, J., Gomez, J. M. & Alexescu, M. (1980). La planète Jupiter en 1975–76. *L'Astronomie*, **91**, 115–29.

Dragesco, J. & McKim, R. (1987). A visit to the Pic-du-Midi Observatory. *J.B.A.A.*, **97**, 280–7.

Eastman Kodak (1972). *Contrast Index, Guide to Proper Development.* Publ. Q-120.

Eastman Kodak (1976). *Practical Processing in Black and White Photography.* Publ. No. P-229.

Eastman Kodak (1977). *Basic Photographic Sensitometry Work Book.* Publ. Z22 ED.

Eastman Kodak (1981). *Filtres Kodak pour usages scientifiques et techniques.* Kodak-Pathé XB-3Fc.

Eastman Kodak (1989a). *Kodak Scientific Imaging Products.* Publ. L-10.

Eastman Kodak (1989b). *Films et plaques pour la photo astronomique, renseignements supplémentaires.* Publ. No. L-16.

Eaton, G. (1965). *Photographic Chemistry.* New York: Morgan & Morgan (Eastman Kodak).

Eggleston, J. (1984). *Sensitometry for Photographers.* London: Focal Press.

Fabry, C. & Arnulf, A. (1937). *La vision dans les Instruments d'Optique.* Paris: Editions Revue d'Optique.

Fielder, G. (1961). *Structure of the Moon's Surface.* Oxford: Pergamon Press.

Fountain, J. W. & Larson, S. M. (1971). Multicolor photography of Jupiter. *LPL Comm.* No. 174.

Gaviola, Z. (1948). On seeing fine structure of stellar images and inversion layer spectra. *Astr. J.*, **54**, 155.

Gerhart, K. (1962). Ein Protuberanzfernrohr für Sternfreunde. *Orion*, **78**, 252–9.

Glafkides, P. (1976). *Chimie et Physique photographiques.* Paris: P. Montel.

Gregory, J. (1957). A Cassegrainian–Maksutov telescope design for the amateur. *Sky and Telescope* (March), 236–9.

Gutschewski, C. L., Kinsler, D. C. & Whitaker, E. (1971). *Atlas and Gazetteer of the Near Side of the Moon.* NASA. (SP-241).

Hansen, T. P. (1970). *Guide to Lunar Orbiter Photographs.* NASA. (SP-242).

Hatfield, H. (1968). *Amateur Astronomers Photographic Lunar Atlas.* London: Lutterworth Press.

Hecquet, J. (1990). Charles Boyer (1911–1989). *Pulsar*, No. 677, 56–8.

Heudier, J. L., Labeyrie, C. & Maury, A. (1981a). Hypersensitizing Kodak Technical Pan 2415. *Astronomical Photography.* Proceedings UAI.

Heudier, J. L., Labeyrie, C. & Maury, A. (1981b). Hypersensitizing Kodak Technical Pan 2415. *Astronomical Photography.* CNRS-INA, 304.

Hill, H. (1991). *A Portfolio of Lunar Drawings.* Cambridge: Cambridge University Press.

Jamieson, H. D. (1986). Getting started: Telescope selection. *Strolling Astronomer*, **35**, 181–3.

Keenan, P. C. (1952). Photography of the sun's disk in integrated light. In Kuiper, G. P., *The Sun.* Chicago: Chicago University Press.

Kopal, Z. (1969). *The Moon.* Dordrecht: Reidel.

Kopal, Z. (1971). *The New Photographic Atlas of the Moon.* New York: Taplinger Publ.

Kopal, Z. (1974). *The Moon in the Post-Apollo Era.* Dordrecht: Reidel.

Kopal, Z. & Garder, R. W. (1974). *Mapping the Moon.* Dordrecht: Reidel.

Kowalaiski, P. (1972). *Théorie photographique appliquée.* Paris: Masson.

Krick, J. (1992). Building Owl Observatory. *Astronomy*, April, May and June issues.

Kuiper, G. (1972). High resolution planetary observation. *Space Res.*, 1683–7.

Kuiper, G. P., Arthur, D. W. G., Moore, E., Tapscott, J. W. & Whitaker, E. A. (1960). *Photographic Lunar Atlas.* Chicago: Chicago University Press.

Kuiper, G. & Middlehurst, B. A. (eds.) (1961). *Planets and Satellites.* Chicago: Chicago University Press.

Kutter, A. (1965). Der Schiefspiegler. *Sterne und Weltraum.* (Jan.), 12–16.

Kutter, A. (1975). A new three-mirror unobstructed reflector. *Sky and Telescope*, **49**, Jan., 46; Feb., 115.

Larson, S. M., Fountain, J. W. & Minton, R. B. (1973). Color photography of Jupiter. *LPL Comm.*, No. 189.

Lille, W. (1992). Shooting the Sun with a backyard telescope. *Sky and Telescope*, **83**, (March), 340–2.

Loughead, R. E., Bray, R. J., Tapperer, E. J. & Winter, J. G. (1968). High resolution photography of the solar chromosphere. *Solar Phys.*, **4**, 185–95.

Lyot, B. (1953). Photographie de Jupiter en 1945. *L'Astronomie*, **67**, 21.

McKim, R. (1993). The life and times of E. M. Antoniadi, 1870–1944. *J.B.A.A.*, **103**, 164–70, 219–27.

McKim, R. & Dragesco, J. (1987). Observations planétaires à l'Observatoire du Pic-du-Midi. *Pulsar*, No. 658, 22–5.

Maksutov, D. D. (1944). New Catadioptric meniscus systems. *J. Opt. Soc. Amer.*, **34**, 270–84.

Malin, D. F. (1977). Unsharp masking. *A.A.S. Photo Bull.*, No. 16, 10–13.

Malin, D. F. & Murdin, P. (1984). *Colours of Stars*. Cambridge: Cambridge University Press.

Martinez, P. (1987). *Astrophotography II*. Richmond, VA: Willmann Bell.

Maunder, E. W. & Maunder, A. S. (1903). Some experiments of limit of vision for lines and spots. *J.B.A.A.*, **13**, 344.

Mazereau, P. & Bourge, P. (1985). *A la poursuite du soleil, la construction du coronographe d'amateur*. Paris: Eyrolles.

Meeus, J. (1983). *Astronomical Tables of the Sun, Moon and Planets*. Richmond, VA: Willmann Bell.

Meinel, A. B. (1960). Astronomical seeing and observatory site selection. In Kuiper, G. & Middlehurst, B. A., *Telescopes*, pp. 158–75. Chicago: Chicago University Press.

Miyamoto, S. and Matsui, M. (1960). Photographic atlas of the Moon. *Contr. Kwasan Obs. Kyoto*.

Mottoni, G. de (1970). Cartografia del Pianeta Marte basata su documentazione fotografice internazionale a partire del 1907, II. Opposizione del 1954 al 1958. *Pub. Osserv. Astr. Milano-Merate, Nuova Serie*, No. 21.

Numazawa, S. (1992). Using a CCD on the planets. *Sky and Telescope*, **83** (Feb.), 209–14.

Paech, W. (1989). Der Protuberenzenansatz von Baader mit dem refraktor Vixen 80/910. *Sterne und Weltraum*.

Page T. & Page, L. W. (1969). *Telescopes*. London: Collier Macmillan.

Paperlein, D. (1969). Des Jahresgang des Nachtmitagsminima der terrestriches Szintillation. *Optik*, **60**, 93–5.

Parker, D. C. (1980). Planetary projection photography: a simple device featuring a vibrationless shutter and seeing monitor. *The Astrograph*, June–July, 85–7.

Parker, D. & Berry, R. (1992). Planetary imaging with a small CCD camera. *Strolling Astronomer*, **36**, 1–9.

Photo-Lab-Index (1979). New York: Morgan & Morgan.

Pickering, W. H. (1895). Artificial disks (visual observations of the moon and planets). *Harv. Ann. (U.S.A.)*, **32**, 117–57.

Pickering, W. H. (1920). Definition and resolution. *Popular Astronomy*, **28**, 510–13.

Pope, T. & Osypowski, T. (1969). High resolution photography. In Page, T. & Page, L. W. (eds.) *Telescopes*, pp. 190–4. London: Collier Macmillan.

Price, W. (1988). *The Moon Observer's Handbook*. Cambridge: Cambridge University Press.

Q. Mike and Pat (1978). *The Manual of Slide Duplicating*. New York: Amphoto.

Reese, E. J. (1970). Jupiter's Red Spot in 1968–69. *Icarus*, **12**, 249–57.

Reese, E. J. (1971). Jupiter: its Red Spot and other features in 1969–1970. *Icarus*, **14**, 343–54.

Reese, E. J. (1972). Jupiter and its Red Spot and disturbances 1970–1971. *Icarus*, **17**, 57–72.

Reese, E. and Solberg, F. G. (1966). Recent measures of the latitudes and longitudes of Jupiter's Red Spot. *Icarus*, **5**, 1266–72.

Reynolds, M. and Parker, D. C. (1988). Hypered film for planetary photography. *Sky and Telescope*, **76** (June).

Roddier, F. (1978). Observations of the sun with interferometry and speckle-interferometry techniques. *Osservazioni e memorie dell'Osservatorio di Arcetri, Florence*, **106**.

Roddier, F. (1981). The effects of atmospheric turbulence in optical astronomy. *Progr. Optics*, **19**, 280–368.

Rösch, J. (1959). Observations de la photosphère solaire. *Ann. Astrophys.*, **22**, 571–83.

Rösch, J. (1990). Charles Boyer (1911–1989) et la rotation de Vénus. *L'Astronomie*, **104**, 216–18.

Rösch, J. & Dragesco, J. (1980). The French quest for high resolution. *Sky and Telescope*, **59** (Jan.), 6–13.

Roth, G. (1960). *Handbook for Planet Observers*. London: Faber & Faber.

Rouvière, F. (1979). Photographie solaire avec un télescope de 20 cm. *L'Astronomie*, **93**, 185–8.

Rükl, A. (1976). *La lune, Venus, Mars*. Paris: Gründ.

Rutten, H. & van Venrooij, M. (1988). *Telescope Optics*. Richmond, VA, Willmann Bell.

Saget, J. (1952). *La photographie astronomique*. Paris: Prisma.

Salanave, L. G. (1957). Observing at Junipero Serra Park. *Sky and Telescope*, **16**, 321.

Schultz, P. (1976). *Moon Morphology*. Austin: University of Texas Press.

Shirao, M. and Sato, S. (1987). *Photographic Guide to the Moon*. Tokyo: Rippu Sholo.

Sidgwick, J. B. (1980). *Amateur Astronomer's Handbook* (Muirden revision). Hillside, NJ: Enslow Publ.

Sigler, R. D. (1974). Family of Compact Schmidt–Cassegrain telescope designs. *Appl. Optics*, **13**, 1765–6.

Sigler, R. D. (1975). Compound Schmidt telescope designs with non-zero Petzval curvatures. *Appl. Optics*, **14**, 2302–3.

155

Sigler, R. D. (1978). Compound catadioptric telescopes with all spherical surfaces. *Appl. Optics*, **17**, 1519–26.

Slipher, E. C. (1962). *The Photographic Story of Mars*. Flagstaff, Arizona: Sky Publishing and Northland Press.

Slipher, E. C. (1964). *A Photographic Study of the Brighter Planets*. Flagstaff, Arizona: Lowell Observatory and National Geographic Society.

Solberg, H. G. (1969). A 3-month oscillation in the longitude of Jupiter's Red Spot. *Planet. Space Sci.*, **17**, 1573–88.

Stix, M. (1989). *The Sun*. Berlin: Springer-Verlag.

Stock, J. and Keller, G. (1961). Astronomical seeing. In Kuiper, G. P. & Middlehurst, B. M. (eds.) *Telescopes*, pp. 138–53, Chicago: Chicago University Press.

Texereau, J. (1984). *How to Make a Telescope*. Richmond, VA: Willmann Bell.

Texereau, J. & Vaucouleurs, G. de (1954). *L'Astrophotographie d'amateur*. Paris: Editions Revue d'Optique.

Thérin, G. (1993). High resolution lunar and planetary photography. In Muirden, J. (ed.), *Sky Watchers Handbook*, pp. 305–29. W. H. Freeman/Spektrum.

Todd, H. N. and Zakia, R. D. (1986). *Photographic Sensitometry*. New York: Morgan & Morgan.

Vaucouleurs, G. de (1958). *La photographie astronomique*. Paris: Albin Michel.

Vaucouleurs, G. de, Dragesco, J. & Selme, P. (1956). *Manuel de photographie scientifique*. Paris: Editions Revue d'Optique.

Vaugh, F. (1969). Astronomical seeing. In Page, T. and Page, L. W. (eds.) *Telescopes*, pp. 129–92. London: Collier Macmillan.

Verseau, R. (1986). Entraînement horaire type secteur lisse d'un télescope (contrôle de la précision du mouvement). *L'Astronomie*, **100**, 299–307*.

Viscardy, G. (1987). *Atlas-guide photographique de la lune*. Paris: Masson.

Walker, G. (1987). *Astronomical Observations*. Cambridge: Cambridge University Press.

Wallis, B. P. & Provin, R. W. (1988). *A Manual of Advanced Celestial Photography*. Cambridge: Cambridge University Press.

Wilkins, H. P. & Moore, P. (1961). *The Moon*. London: Faber & Faber.

Willey, R. R. (1962). Cassegrain type telescopes. *Sky and Telescope*, **23**, 226.

Woods, N. Del (1977). Solar observations with narrow band interference filters. In Maag, R. C., Sharlin J. M. & Van Zandt, R. (eds.) *Observe and understand the Sun*, pp. 27–9. Washington: Astronomical League.

Zirin, H. (1988). *Astrophysics of the Sun*. Cambridge: Cambridge University Press.

Zmek, W. P. (1993). Rules of thumb for planetary scopes I and II. *Sky and Telescope*, **86**, July, 91–5; Sept., 83–7.f

---

* For an English version, see: Verseau, R. (1994). A smooth sector lead-screw drive system. *J.B.A.A.* **104**, 222–4.

Index